HYDROCARBON POLYMER ELECTROLYTES FOR FUEL CELL APPLICATIONS

HYDROCARBON POLYMER ELECTROLYTES FOR FUEL CELL APPLICATIONS

JINLI QIAO
AND
TATSUHIRO OKADA

Nova Science Publishers, Inc.
New York

LIBRARY OF CONGRESS CATALOGING-IN-PUBLICATION DATA
Hydrocarbon polymer electrolytes for fuel cell applications / Jinli Qiao and Tatsuhiro Okada (editor).
    p. cm.
 ISBN 978-1-60456-846-2 (softcover)
 1. Conducting polymers. 2. Polyelectrolytes. 3. Hydrocarbons. 4. Fuel cells. I. Qiao, Jinli. II. Okada, Tatsuhiro.
QD382.C66H93   2008
621.31'2429--dc22   2008023082

Published by Nova Science Publishers, Inc. + *New York*

CONTENTS

PREFACE

Direct methanol fuel cells (DMFCs), employing liquid methanol as a fuel, offer an attractive option in portable devices due to their simplicity in the system structure (easy storage and supply), no need for fuel reforming or humidification. For obtaining a higher power density, the membranes that show high proton conductivity, and at the same time, low methanol permeability are strongly desired. However, there is achieved only a little progress because of trade-off relations between these parameters. Also the membrane stability, particular to hydrolytic and chemical stability is recognized as a key factor that affects fuel cell performances. In our recent work, we have been working on the design and the development of new families of cost-effective, readily prepared proton-conducting membranes based on chemically cross-linked PVA-PAMPS [poly(vinyl alcohol) and poly(2-acrylamido-2-methyl-1-propanesulfonic acid)] composites. We have first introduced new concepts of secondary polymer chains such as "binary chemical cross-linking" or "hydrophobicizer" and the "stabilizer" effect. Also, we establish a new concept of PVA-PAMPS based semi-interpenetrating polymer networks (semi-IPNs) by incorporating plasticizer variants R (R = poly(ethylene glycol)(PEG), poly(ethylene glycol) methyl ether (PEGME), poly(ethylene glycol) dimethyl ether (PEGDE), poly(ethylene glycol) diglycidyl ether (PEGDCE)) and poly(ethylene glycol)bis(carboxymethyl)ether (PEGBCME) as the third components. Incorporation of the above concepts promoted not only the high proton conductivity (~ 0.1 S cm^{-1}), flexibility with low methanol permeability ($1/3 - 1/2$ of Nafion 117 membrane), but also the excellent hydrolytic and the oxidative stability of PVA-PAMPS composites.

The membrane electrode assembly (MEA) fabricated with PVA-PAMPS composites has been successfully established, which showed the similar open circuit voltage (OCV) to that of Nafion 115, and a power density ~ 52 mW cm^{-2} at

80°C. A striking feature of the long-term test was that no appreciable decay of the current density was observed during the whole operation time longer than130 hours at 50°C, and so was the power density. This is the first time that such long-term operation of DMFC was reported since PVA-PAMPS composite are all hydrocarbon membranes made simply of aliphatic skeletons. They are very different from the perfluorosulfonic membranes such as Nafion[®], or other reported membranes with aromatic skeletons. Therefore this affords the PVA-PAMPS composites unique structure compared to most of the proposed membranes, which suggests the good candidacy of PVA-PAMPS composites when they are intended for use in low temperature DMFCs.

Chapter 1

INTRODUCTION

Fuel cells are clean alternative to current technologies, which transform chemical energy directly into electricity. Because of no circumvent Carnot cycle limitations, the fuel cells can, in principle, supply energy with efficiencies in excess of 80% and, this technology has been expected to be the most promising environmentally-friendly power sources in road transportation and stationary applications [1]. As a key component, polymer electrolyte fuel cells (PEFCs) utilize a proton-conducting polymer membrane as solid electrolyte, therefore, the corrosion problem of the cell is highly overcome.

The proton-exchange membrane fuel cells (PEMFCs) are mainly divided into two sorts, which are called $H_2/O_2(H_2/air)$ fuel cells and direct methanol fuel cells (DMFCs), respectively, depending on their different fuels used. For a H_2/O_2 fuel cell or simply called polymer electrolyte fuel cells (PEFCs), the electrochemical reaction occurring on the anode is a 2-electron transfer hydrogen oxidation reaction (Schematic 1), therefore, high power density, rapid start-up and high efficiency could be obtained. In DMFCs, on the other hand, the fuel methanol is oxidized via 6-electron transfer process. It, therefore, results in a slow kinetics of the methanol oxidation on the anode catalyst surface. It is to be noted that PEFCs using pure hydrogen (H_2) as fuel will involve unsolved technological problems and economic uncertainties either on-site H_2 storage (compact and lightweight hydrogen storage) or an onboard reformer to extract H_2 from organic fuels (hydrogen supply, and distribution and refueling systems).

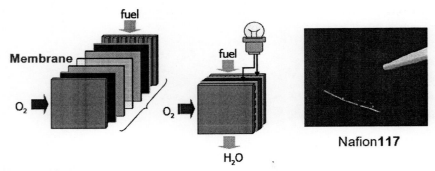

when H_2 as a fuel:

Anode: $H_2 \rightarrow 2H^+ + 2e$, E = 0.0V   (1)

Cathode: $2H^+ + 2e + 1/2O_2 \rightarrow H_2O$, E = 1.229V   (2)

when CH_3OH as a fuel:

Anode: $CH_3OH + H_2O \rightarrow CO_2 + 6H^+ + 6e$, E = 0.046V   (3)

Cathode: $3/2O_2 + 6H^+ + 6e \rightarrow 3H_2O$, E = 1.229V   (4)

As an alternative, on the other hand, DMFCs employing liquid methanol as a fuel offer an attractive option in portable devices such as cellular phones, personal digital assistants (PDAs) and laptop computers for civilian and defense applications [2–4] due to their high energy density, simplicity in the system structure (easy storage and supply), no need for fuel reforming or humidification. This system simplification gives DMFCs an advantage over PEFCs with reformed H_2 fuel. It also gives perspective to replace even the most advanced rechargeable batteries (e.g. nickel–metal hydride and lithium ion) currently used for such applications [5-7].

Whether it is for PEFCs or for DMFCs, high proton conductivity is a key factor for proton exchange membranes (PEMs) since it allows for higher performances to be achieved. Nafion®, the perfluorosulfonic acid (PFSA) membranes, developed by DuPont, is the common denominator in this technology operating at a typical temperature of 80°C. They are composed of carbon-fluorine backbone chains with perfluoroether side chains containing sulfonic acid (figure 1). This Teflon-like molecular backbone gives these materials excellent long-term stability in both oxidative and reductive environments. Unfortunately, their high cost and some unsolved technological problems such as related to high methanol crossover and difficulty in synthesis and processing still delays the commercial production. Especially in DMFC, a large methanol crossover rate of about 10^{-6} mol cm^{-2} s^{-1} for PFSA membranes corresponds to a performance loss of current density of 50-100 mA cm^{-2} [5]. This results in not only the loss of fuel, but also the performance degradation at the cathode due to the mixed reaction of methanol oxidation with oxygen reduction reaction [8]. Furthermore, the platinum catalyst used in the cathode can be easily poisoned by CO-based compounds generated in the oxidation of methanol, resulting in constant cathode depolarization [9]. It leads to a high anodic over-potential loss of ca. 350 mV, compared to ca. 60 mV for hydrogen [10]. Therefore, a successful DMFC membrane should be not only conduct protons, but also prevent methanol crossover.

The book attempts to summarize recent significant progress in synthesis of novel PEMs with the help of the results obtained in the author's laboratory or reported in the literatures. Less attention is given to the nonperfluorosulfonated membranes, particular to the sulfonation and cross-linking of polyaromatic and polyheterocyclic polymers since these subjects have been well-treated in recent reviews [1,10,11-16]. Searching for better membrane materials such as developing effective membrane systems with low cost, easy preparation and simple structure and improving chemical and structural stability of the current membrane materials are the motivation of these developments to substitute PFSA membranes for low temperature DMFC operation (~80°C). The poly(vinyl alcohol) (PVA) based PEMs are reported in emphasis because of their superior methanol barrier advantage. The latest developments on alkaline fuel cell membranes using modified PVA at low, medium temperature are also reported.

$$—(CF_2—CF_2)_x—(CF_2—CF)_y—$$
$$(O—CF_2—CF)_m—O—(CF_2)_n—SO_3H$$
$$CF_3$$

$$m = 1, x = 5\text{-}13.5;$$
$$n = 2, y = 1$$

Figure 1. Structure of Nafion® membranes.

Chapter 2

PVA-PAMPS AS NOVEL HYDROCARBON PROTON-CONDUCTING ELECTROLYTE MEMBRANES

Commercial poly(vinyl alcohol) (PVA) is a polyhydroxy polymer, which is very common in practical applications because of its easy preparation and biodegradability [16]. It is usually derived from poly(vinyl acetate), which are mainly used for paper and textile sizing, oxygen resistant films, adhesives, food wrappings, and desalination and pervaporation membranes [17-19]. Owing to its high selectivity of water to alcohols, PVA membranes have been used in alcohol dehydration to break the alcohol–water azeotrope [20-23]. Perceiving the reasonable selectivity of water to alcohols in the commercial PVA membranes, Pivovar et al. [24] explored the potentiality of PVA as proton exchange membrane in DMFC based on proton conductivity and methanol permeability. It was found that the PVA membranes employed in pervaporation process were much better methanol barriers than Nafion® membrane. However, the PVA membranes are the poor proton conductors as compared with Nafion® membrane since the PVA itself does not have any negative charged ions such as carboxylic or sulfonic acid groups, while for efficient fuel cells, the PEMs must be very resistant to the oxidation and have conductivity $\geq 10^{-2}$ S cm^{-1} [25].

Progress has been made in our lab with a new polymer, poly(2-acrylamido-2-methyl-1-propanesulfonic acid) (PAMPS) as an novel proton conductor [26-33]. PAMPS was found to be a high proton conductor comparable to partially hydrated Nafion® due to sulfonic acid groups in its chemical structure [34]. The application of PAMPS can be found in electrochromic devices as a proton conducting gel [35] and humidity sensor [36]. The copolymers of PAMPS with other monomers used as PEMs in fuel cells are also reported [34, 37]. It showed a value of σ as high as 0.2 S cm^{-1} at 70 °C for a copolymer containing 54 mol% AMPS equilibrated at 98%RH [34].

In fact, PAMPS could be directly dispersed into the PVA net-work by a simple solution casting method [26-33]. The membrane structure was finished by cross-linking the hydroxyl groups of PVA main chain with acetal ring formation by immersing the membranes in a reaction agent, where glutaraldehyde (GA), for example, was used as a cross-linker. The cross-linking reaction is finished just in 30-60min. Since the whole process proceeds just at room temperature, any possible interference due to the complicated organic synthesis procedures is avoided. Thus, it realized a real concept on "low cost, simple preparation and easy fabrication". In addition, PAMPS used is not simply as a proton conductor but also a catalyst because the sulfonic groups (-SO$_3$H) in PAMPS is sufficient for the acid-catalyzed cross-linking reactions. It, therefore, does not need acid catalyzers like HCl or H$_2$SO$_4$ by further addition from outside [38-40]. Not limited to this, since PAMPS is a polymer with high weight molecule of 2, 000, 000, it can be well trapped in the PVA net work by strong hydrogen-bonding and, does not leak out from the resulted membranes, which is often an indicted problem for smaller inorganic acids doped ones [41].

Chapter 3

PLASTICIZER INCORPORATED PVA-PAMPS SEMI-INTERPENETRATING POLYMER NETWORKS

2.1.1. MEMBRANE PREPARATION AND MECHANICAL PROPERTIES

PVA-PAMPS proton-conducting semi-interpenetrating polymer networks (semi-IPNs) could be prepared easily by a simple solution casting method, where PVA (average Mw = 124,000–186,000) was fully dissolved in water to make a 6% solution at 70–80°C. PAMPS (average Mw = 2,000,000) and plasticizers R (R: polyethylene glycol (PEG), poly(ethylene glycol) methyl ether (PEGME), poly(ethylene glycol) dimethyl ether (PEGDE) and poly(ethylene glycol) diglycidyl ether (PEGDCE), Mw = 400 – 600) were separately prepared and mixed to cast a membrane at ambient temperatures. After soaking the membrane square pieces in a reaction solution containing 10 mass % GA in acetone at 25°C, the cross-linking proceeded between the hydroxyl groups (-OH) of PVA and the aldehyde groups (-CHO) of GA in the membrane due to an acid-catalyzed reaction by PAMPS. Figure 2 illustrates schematic structures of PVA-PAMPS-R semi-IPNs and the membrane pictures. The membrane structure was demonstrated by a FTIR result shown in figure 3. Transparent, flat membranes were obtained with a thickness of about several tens of micrometers (60–100μm). The thickness of the membranes could be controlled easily by adjusting the volume of suspension.

Note when plasticizer variants PEG, PEGME, PEGDE and PEGDCE as mentioned above are incorporated into PVA/PAMPS semi-IPNs, the resulted membranes are found to improve so much in the screening experiment [29-31,33]. The membranes would become too brittle to endure even a slightly hard handling if the membrane was prepared merely from PVA and PAMPS components, while the ternary-composed PVA-PAMPS are robust and flexible. The membranes exhibit good mechanical property when plasticizer, such as PEGDCE content ranges 11 – 30 mass%, PEG content ranges 11 – 35 mass%, PEGME content ranges 13 – 35 mass% and PEGDE content ranges 11 – 27 mass%, respectively. Beyond these ranges, either the brittle membrane or obvious phase separation occurs. PEG has been reported to have a unique ability to be soluble in both aqueous and organic solvents due to its hydrophilic and hydrophobic moieties and, is miscible easily with other polymers such as PVA [42–44]. Since the addition of plasticizers would assist the relaxation of the polymer chains through their special plasticizing effect, it may lead to a good dispersion of phase separation in these membranes [29-31, 33], thus the micro-structure of the membranes is highly improved by incorporating plasticizer variants. The special plasticizer effect will also influence the membrane stability which is also a key issue for a practical fuel cell usage and will be discussed in the following sections.

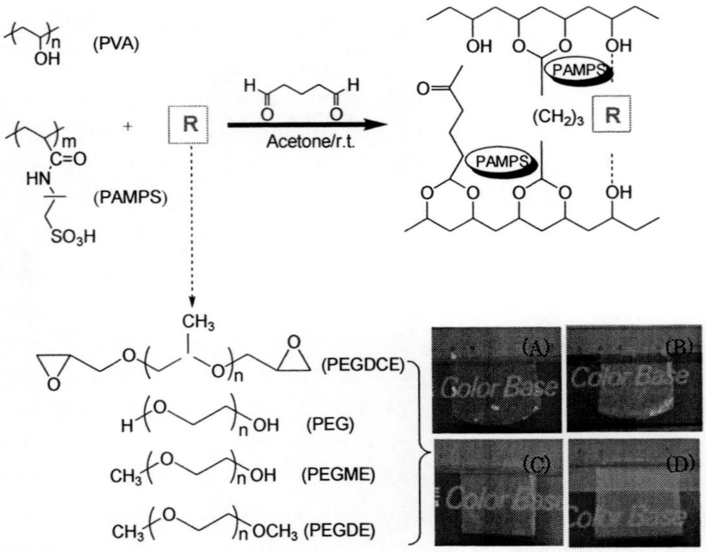

Figure 2. Chemical structure of PVA-PAMPS-R semi-IPNs. R = (A) PEGDCE, (B) PEG, (C) PEGME, and (D) PEGDE, respectively.

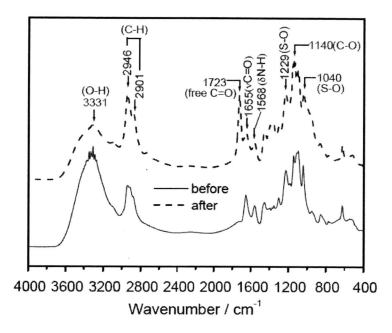

Figure 3. FT-IR spectra of PVA-PAMPS-R semi-IPNs. Polymer composition:
PVA/PAMPS/R = 1 :1 :0.75 in mass. R = PEGDCE.

2.1.2 THERMAL PROPERTY

The plasticizer incorporated PVA-PAMPS semi-IPNs have a three-step thermal decomposition (figure 4). The first step occurred at around a temperature range 80–150°C, where water desorption occurred (10%). Then through 190–330°C, PAMPS (sulfonic acid) and PVA decomposed with a weight loss of 45%, followed by a major degradation of cross-linking bridge and polymer backbone through 330–500°C. For PVA alone, very large decomposition was observed between 330 and 500°C, and more than 80% of the polymer decomposed almost completely. Therefore, they are well above the intended fuel cell use temperature range (ca. 30 – 150°C). Glass transition temperature (T_g) was not obviously observed or lower than pristine PVA for the above semi-IPNs. This may be due to a complex composition by cross-linking in the membranes which is very different from other pure systems such as erfluorosulfonated membranes.

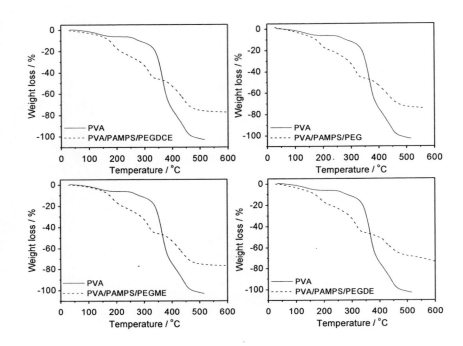

Figure 4. TG profiles of PVA-PAMPS-R semi-IPNs. Polymer composition: PVA/PAMPS/PEGDCE or PEG =1:1:0.5, PVA/PAMPS/PEGME = 1:1:0.4, and PVA/PAMPS/PEGDE = 1:1:0.3 in mass, respectively. Heating rate: 5°C min⁻¹.

2.1.3. SWELLING BEHAVIOR AND PROTON CONDUCTIVITY

The presence of water in the membranes is a prerequisite for reaching high proton conductivity. Especially in PEM fuel cells, higher levels of proton conductivity allow for higher power densities to be achieved. On the other hand, excessively high water uptake will affect other performances of the membranes such as the dimensional and thermal stability. Shown in figure 5 is the water uptake (WU), expressed as grams of water incorporated per gram of dry membrane and, this was evaluated as a measure of cross-linking density. It can be seen that the cross-linking density of the membranes strongly depended on the cross-linking time. It increased with cross-linking time accompanied by an improvement in mechanical property of the membranes and then a small decrease in proton conductivity due to the reduced water absorptions.

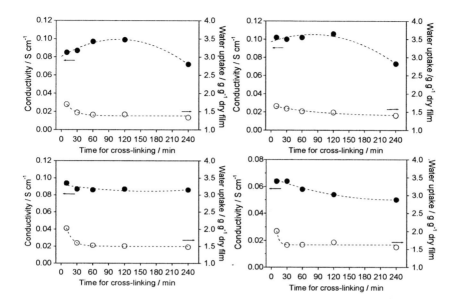

Figure 5. Change of water uptake and proton conductivity of PVA-PAMPS-R semi-IPNs with cross-linking time. Polymer composition: PVA/PAMPS/R = 1:1:0.5 in mass. R = (A) PEGDCE, (B) PEG, (C) PEGME and (D) PEGDE, respectively.

The WU and proton conductivity of PVA-PAMPS-R semi-IPNs are also strongly depended on humidity, content of plasticizer and PAMPS incorporated. Figure 6 shows this relation of PVA-PAMPS semi-IPNs by applying different plasticizer variants. The equilibrium water uptake of the membranes increased with the content of plasticizer due to their strong hydrophilicity, where a decrease in proton conductivity occurred probably because of a diluting effect of charge carriers (figure 6A-6D). In every case, the proton conductivities measured at 25°C with ac impedance spectroscopy reached as high as 0.1 S cm^{-1}, regardless of plasticizer variants but with different polymer compositions (table 1). By changing the content of PAMPS in the membranes, the water uptake and proton conductivity both increased due to increased sulfonic acid groups, and shows a linear relationship by incorporating PEGDCE and PEG plasticizers (figure 6A' and 6B'). Contrary to this, applying PEGME and PEGDE as plasticizers lead to an initial increase in proton conductivity then leveled off and, even experienced a decrease with further increasing PAMPS content in the membranes (figure 6C' and 6D'). This is considered to be a diluted effect larger than the effect of charge carriers by even further increased sulfonic acid groups. When the mass ratio of

PVA/PAMPS was above 1:1, the membrane became brittle, therefore, the mass ratio of PVA/PAMPS/PEGDCE or PEG = 1:1:0.5 in mass, PVA/PAMPS/PEGME = 1:1:0.4 in mass and PVA/PAMPS/PEGDE = 1:1:0.3 in mass is supposed to be the optimized value with both high proton conductivity and good flexibility (table 1).

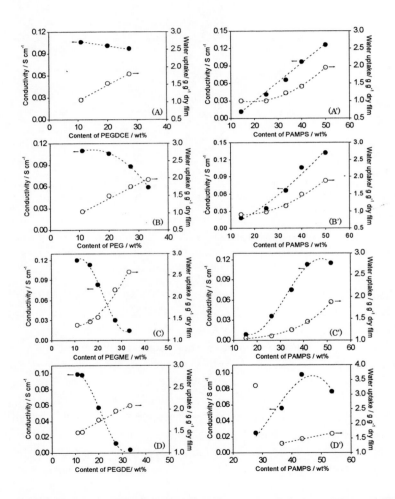

Figure 6. Change of water uptake and proton conductivity of PVA-PAMPS-R semi-IPNs in different content of R and PAMPS at 25°C. Polymer composition: (A), (B), (C), and (D) PVA/PAMPS = 1:1 in mass; (A') PVA/PEDCE = 1:0.5, (B') PVA/PEG = 1:0.5, (C') PVA/PEGME = 1:0.4 and (D') PVA/PEGDCE = 1:0.3 in mass, respectively.

Table 1. Proton conductivity (σ), ion exchange capacity (IEC), water uptake (WU), methanol permeability (P_M) and selectivity (σ/P_M) of PVA-PAMPS-R semi-IPNs. [a]

R	Polymer composition (PVA/PAMPS/R)	Time for cross-linking (min)	Membrane thickness (μm)	σ S cm^{-1}	IEC mequiv. g^{-1}	WU (gg^{-1} dry polymer)	P_M ($\times 10^6$ cm^2 s^{-1})	σ/P_M ($\times 10^4$ S cm^{-3} s)
PEGDCE	1:1:0.5	60	60-100	0.099	1.78	1.41	1.01	9.80
PEG	1:1:0.5	60	60-100	0.109	1.85	1.49	1.17	9.06
PEGME	1:1:0.4	60	60-100	0.113	1.96	1.43	1.02	11.8
PEGDE	1:1:0.3	40	60-100	0.099	1.43	1.48	0.96	10.3
PEGBCME	1 :0.75 :0.4	60	60-100	0.095	1.79	0.91	0.78	12.2
Nafion 117			185	0.1	0.92	0.34	2.13	4.69

[a]10 mass% GA as a cross-linking agent, T = 24 ± 1°C.

2.1.4. THE ION EXCHANGE CAPACITY (IEC), CONCENTRATION (C$_{SO3-}$) AND MOBILITY (U$_{SO3-}$) OF CHARGE CARRIES

IEC is known to have the profound effect on membrane conductivity except for WU. On the one hand, the proton conductivity increases because of increased mobility of charge carriers (ions) in the water phase with increasing water content (volume), on the other hand, the proton conductivity increases with increasing IEC because of the high charge density of the membranes. From figure 7, we can see that the IEC increased almost linearly with the content of PAMPS in the membranes (by applying PEGDCE and PEG as plasticizsers) and then leveled off (by applying PEGME and PEGDE as plasticizers) but decreased simply in all cases with the content of plasticizer. This coincides quite well with the proton conductivity and water uptake as figure 6 shows. A large PAMPS content plays a major role in controlling the proton conduction due to the increased sulfonic acid groups in the membrane, while a larger increase in the amount of water due to incorporated plasticizer variants does not simply make an additional contribution to the conductivity but rather a dilution of charge carriers. In general, the IEC for PVA-PAMPS semi-IPNs ranged from 1.43 mequiv g−1 by applying PEGDE as plasticizer to 1.96 mequiv g−1 by applying PEGME as plasticizer when PVA/PAMPS = 1:1 in mass, while Nafion 117 exhibited a lower IEC value of 0.92 mequiv g−1 (table 1).

Table 2 gives another interesting parameter for describing the proton conductive behavior of the membranes that is the concentration (C$_{SO3-}$) and mobility (u$_{SO3-}$) of charge carries for PVA-PAMPS-R semi-IPNs together with those for Nafion®[45]. It can be seen that for both Nafion 117 and Nafion 115, they showed a higher concentration of charge carriers but a lower mobility in comparison to plasticizer incorporated PVA-PAMPS, which showed a lower concentration of charge carriers due to a higher water uptake in the membrane but a higher mobility. Therefore, the plasticizer incorporated PVA-PAMPS semi-IPNs would be promising candidates for PEM fuel cells if their water swelling behavior could be suppressed further.

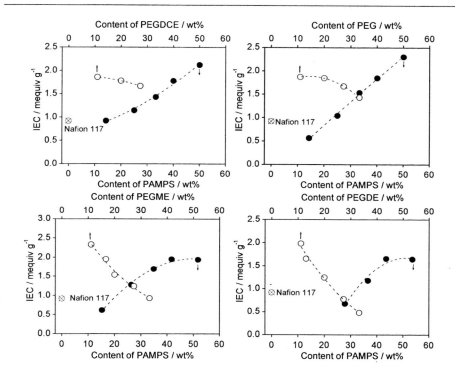

Figure 7. Change of IEC of PVA-PAMPS-R semi-IPNs in different content of R and PAMPS at 25oC. Polymer composition: PVA/PAMPS = 1:1 in mass for dashed mark; (A) PVA/PEDCE = 1:0.5, (B) PVA/PEG = 1:0.5, (C) PVA/PEGME = 1:0.4 and (D) PVA/PEGDCE = 1:0.3 in mass for solid mark, respectively.

Table 2. The concentration (CSO3-) and mobility (uSO3-) of charge carries of PVA-PAMPS-R semi-IPNs together with those of Nafion®

Membrane	Polymer composition	CSO3-× 103 (mol cm-3)	uSO3- × 103 (cm2 V-1s-1)
Nafion 117	–	1.45	0.72
Nafion 115	–	1.17	0.88
PVA-PAMPS-PEGDCE	1 :1 :0.5	0.85	1.26
PVA-PAMPS-PEG	1 :1 :0.5	0.80	1.29
PVA-PAMPS-PEGME	1 :1 :0.4	0.85	1.38
PVA-PAMPS-PEGDE	1 :1 :0.3	0.67	1.53
PVA-PAMPS-PEGBCME	1 :0.75 :0.4	0.92	1.07

2.1.5. TEMPERATURE DEPENDENCE OF PROTON CONDUCTIVITY

Figure 8 shows the proton conductivity as a function of temperature for plasticizer incorporated PVA-PAMPS semi-IPNs in the premium compositions from table 1. All the membrane samples exhibit positive temperature–conductivity dependencies in the tested temperature range from 5 up to 50 oC regardless of plasticizer variants and PAMPS contents in the polymer. This suggests a thermally activated process. The membranes show the highest proton conductivity of 0.15 S cm-1 at 45±2 oC or 0.079 S cm-1 at 5±2 oC depending on different plasticizer variants. The activation energy values Ea fall into the range of 8.6 – 12.3 kJ mol-1 derived from the slope of log $\sigma \sim 1/T$ plots (table 3). These are even a little smaller than that of Nafion 117, for which Ea = 14.6 kJ mol-1 in its fully hydrated state under the same measuring conditions. Because of the amide groups (–C–NH) of PAMPS and the carbonyl groups (C=O) of the cross-linker, GA, after chemical cross-linking (figure 1 and 2), it is reasonable to assume that the proton conduction occurs by two routes, that is, the protons transport via acid groups and amide groups of the polymer molecules through hydrogen bonding. It may be noted that the lowest Ea value was obtained by applying PEGDCE as the plasticizer, and it follows the order PEGCE < PEG < PEGME < PEGDE. Obviously the different structure of the plasticizer gives its different properties and that is why we obtain the similar proton conductivities but in different polymer compositions.

Table 3. Apparent activation energy, Ea, of PVA-PAMPS-R semi-IPNs compared with that of Nafion 117

Membrane	Polymer composition	Ea (kJ mol-1)
Nafion 117	–	14.6
PVA-PAMPS-PEGDCE	1 :1 :0.5	8.64
PVA-PAMPS-PEG	1 :1 :0.5	9.74
PVA-PAMPS-PEGME	1 :1 :0.4	11.18
PVA-PAMPS-PEGDE	1 :1 :0.3	12.32

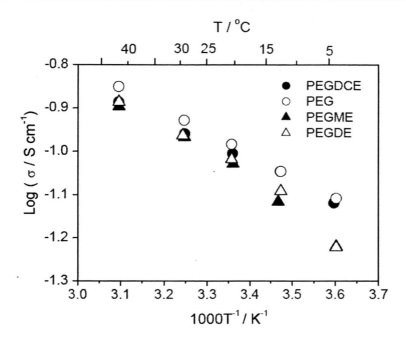

Figure 8. Temperature dependence of proton conductivity for different plasticizer incorporated PVA-PAMPS semi-IPNs.

2.1.6. METHANOL PERMEABILITY (PM) AND SELECTIVITY (σ/PM)

Figure 9 shows the methanol permeability of PVA-PAMPS semi-IPNs, which are plotted as a function of both PAMPS and plasticizer content in the membranes. For a comparison, methanol permeability of Nafion117 was measured under the same experimental conditions. The methanol permeability of Nafion 117 reported as 1.7×10^{-6} cm2 s-1 at room temperature [46] is in good agreement with the present value of 1.83×10^{-6} cm2 s-1 at 24±1oC. The methanol permeability of semi-IPNs increased with PAMPS content, and also increased with additional high content of plasticizer owing to the hydrophilic property of both PAMPS and plasticizer variants. In brief, the methanol permeability was found to be proportional to the proton conductivity. It is important to note that PVA-PAMPS-R semi-IPNs exhibited higher water uptake value in comparison with Nafion 117 (four times higher than that of Nafion 117), but the opposite trend was observed for methanol permeability as table 1 shows. In fact, the semi-IPNs with 51 mass% of PAMPS exhibits still lower methanol permeability compared to Nafion 117, although the water uptake of PVA-PAMPS semi-IPNs was more than five times higher (figure 6). Similar results was also observed in BPSH co-polymer with 50% degree of disulfonation [47].

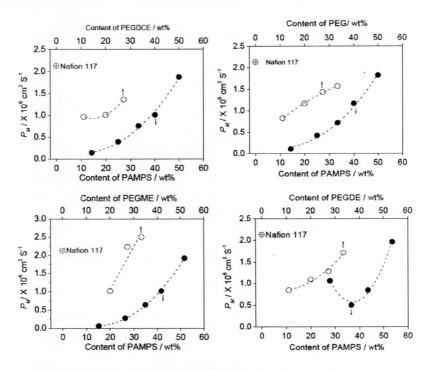

Figure 9. Change of methanol permeability of PVA-PAMPS-R semi-IPNs in different content of R and PAMPS at 25oC. Polymer composition: PVA/PAMPS = 1:1 in mass for dashed mark and, (A) PVA/PEDCE = 1:0.5, (B) PVA/PEG = 1:0.5, (C) PVA/PEGME = 1:0.4 and (D) PVA/PEGDCE = 1:0.3 in mass for solid mark, respectively.

To achieve a successful operation of DMFC, for example, the membrane should possess simultaneously high proton conductivity and low methanol permeability. That is, the ratio of proton conductivity and methanol permeability/or selectivity (σ/PM) is, to some extent, more important than the pure methanol permeability. The higher the selectivity value is, the better performance of DMFC one can expect. From table 1 and figure 10, it can be noted that the selectivity of PVA-PAMPS-R semi-IPNs are all higher than that of

Nafion 117 and, as a result about 2 to 2.5 fold higher than Nafion 117 in their optimized composition.

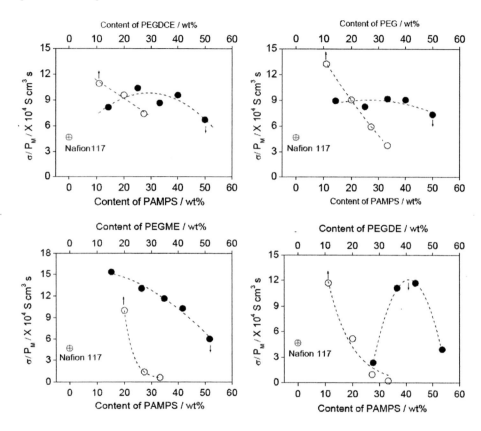

Figure 10. Change of selectivity (σ/PM) of PVA-PAMPS-R semi-IPNs in different content of R and PAMPS at 25oC. Polymer composition: PVA/PAMPS = 1:1 in mass for dashed mark and, (A) PVA/PEDCE = 1:0.5, (B) PVA/PEG = 1:0.5, (C) PVA/PEGME = 1:0.4 and (D) PVA/PEGDCE = 1:0.3 in mass for solid mark, respectively.

As described previously, the microstructure of the membranes may be highly improved by incorporating plasticizer variants. The presence of hydrophilic and hydrophobic moieties in these plasticizer variants gives their unique ability to be soluble in both aqueous and organic solvents, and brings about a good dispersion of phase-separation in these membranes. One more interesting aspect should be addressed to the chemical structure of the polymer, where no aromatic or heterocyclic rings are incorporated to the main chain and the side chains. This is unique structure compared to most of the proposed membranes for DMFC applications [1,11,12-16].

In the recent work, poly(ethylene glycol)bis(carboxymethyl)ether (PEGBCME) that have bifunctional carboxylic acid end groups (figure 11) has been successfully proposed as a novel plasticizer[33]. It was found that PEGBCME could improve the membrane swelling behavior while still afford the membrane high proton conductivity. PEGBCME-incorporated PVA-PAMPS showed a proton conductivity value around 0.095 S cm-1 at 25oC for a polymer composition PVA/PAMPS/PEGBCME = 1:0.75:0.4 in mass, although the content of PAMPS in the membranes is 25% decreased in comparison with other plasticizer incorporated semi-IPNs (PVA:PAMPS = 1:1 in mass) (table 1). The water uptake of the membranes, on the other hand, is suppressed (decreased 33%, table 1) resulting from further cross-linking between the double carboxylic acid end groups (–COOH) in PEGBCME with hydroxyl groups (–OH) in PVA along with strong hydrogen-bonding except for the pristine cross-linking from GA and – OH of PVA [33]. It is demonstrated by the ion-exchange capacity (IEC) of 1.79 for PEGBCME plasticized PVA-PAMPS, which is comparable to other plasticizer incorporated semi-IPNs but with a mass ratio PVA/PAMPS = 1:1 in mass (table. 2).

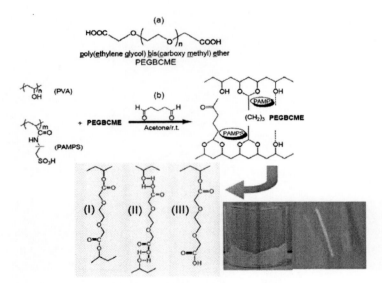

Figure 11. (A) Chemical structure of novel plasticizer PEGBCME, (B) PVA-PAMPS-PEGBCME semi-IPN. Polymer composition PVA/PAMPS/PEGBCME = 1:0.75:0.4 in mass.

2.1.7. STABILITY OF PLASTICIZER INCORPORATED PVA-PAMPS SEMI-IPNS

As we known, for DMFC the membranes that show high proton conductivity and, at the same time, low methanol permeability are strongly desired, and much effort has been undertaken for the above purpose. Howerer, the problem of "membrane stability", particularly hydrolytic and chemical stability, is still a big barrier, and is also recognized as a key factor that affects fuel cell performances [48]. Unfortunately, a little progress is achieved [46, 49-51] because of trade-off relations between these parameters and, the reported data are mainly as to hydrolytic stability [39,46,50]. Different from above situation, PVA-PAMPS-R semi-IPNs are found to exhibit both the good hydrolytic and the oxidative stability [30]. Shown in figure 12 are the results that the hydrolytic stability was tracked by immersing PVA-PAMPS-R semi-IPNs into 0.1 M H2SO4 solution at 60oC. It can be seen that PVA-PAMPS-R semi-IPNs retained high proton conductivity even after soaking in 0.1 M H2SO4 solution for more than 2000 hours. No significant changes were observed in proton conductivity before and after the membranes immersion except for the initial days where high proton conductivities excess to 0.1 S cm-2 may be due to the immersed H+ from H2SO4 solution. These observations indicate that PAMPS is well trapped within the PVA network structure and, the acetalization of main chain is highly resistant to the acid decomposition.

The longevity of the membranes is tested by subjecting to Fenton reagent in an aqueous solution of H2O2(3%)/FeSO4(2ppm). The membrane stability toward the oxidative effect was tracked as weight loss as a function of immersion time at a fixed treatment temperature of 60 °C (figure 13). No weight loss was observed within initial 2-3 hours for these semi-IPNs, then a sharp decrease in the membrane weight occurred within another 2-3 hours, but after that a weight about 70-80 wt% of the originals was still retained even after 30 hours of the treatment. The membranes were able to withstand the solution at least for one-week before they completely dissolved into Fenton's reagent at 60oC. This is a very high oxidative stability seeing that they have no any fluorinated hydrocarbon skeleton and, they are even superior to other hydrocarbon membranes with aromatic skeletons, where the membranes were just subjected to deionised water (80oC) [46], or dissolved into Fenton's reagent within 40 min ~ 2h at 80oC [49].

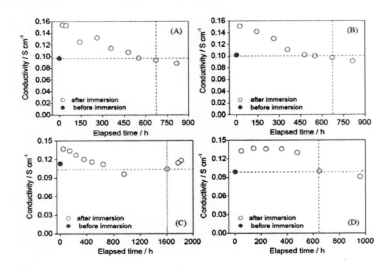

Figure 12. Hydrolytic stability of PVA-PAMPS-R semi-IPNs in 0.1M H2SO4 at 60oC. R
= (A) PEGDCE, (B) PEG, (C) PEGME and (D) PEGDE. Polymer composition
PVA/PAMPS/PEGDCE = 1:1:0.5, PVA/PAMPS/PEG = 1:1:0.5, PVA/PAMPS/PEGME =
1:1:0.4 and PVA/PAMPS/PEGDE = 1:1:0.3 in mass, respectively.

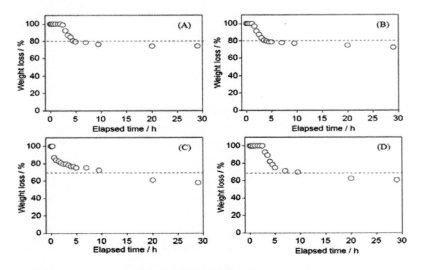

Figure 13.. Oxidative stability of PVA-PAMPS-R emi-IPNs in H2O2(3%)/FeSO4(2ppm)
at 60oC. R = (A) PEGDCE, (B) PEG, (C) PEGME and (D) PEGDE. Polymer composition:
PVA/PAMPS/PEGDCE = 1:1:0.5, PVA/PAMPS/PEG = 1:1:0.5, PVA/PAMPS/PEGME =
1:1:0.4 and PVA/PAMPS/PEGDE = 1:1:0.3 in mass, respectively.

Chapter 4

PVA-PAMPS COMPOSITES ON THE BASIS OF BINARY CHEMICAL CROSS-LINKING/OR SIDE CHAINS

PVA-PAMPS membranes can not be formed before chemical cross-linking. All of the samples would be dissolved in water and converted from a clear, dry membrane to a cloudy, gel-like state just prior to dissolution. Cross-linking is an effective means and several research groups have pursued successfully in their membrane preparations [38-40, 50-52]. Unfortunately, the usual chemical cross-linking technique, that is, using GA as a cross-linker and HCl as a catalyst with acetone or an aqueous solution as a reaction medium [17,38, 53] was found to be not suitable PVA-PAMPS membranes except for by adding plasticizer variants as we proposed in section 2.1. The final membranes obtained are too brittle to endure even a small rigid handling regardless of the cross-linking time or the concentration of the cross-linker.

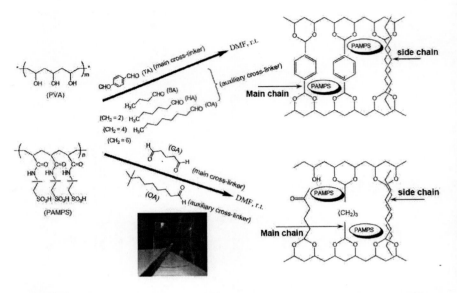

Figure 14. Molecular structures of the cross-linkers used and chemically cross-linked PVA based on a binary chemical cross-linking. Here monofunctional polyalkylenes (two, four, and six CH2 groups) terminated by aldehydes were used as auxiliary cross-linkers to form the side chains, and bifunctional terephthalaldehyde was used to form the main chain.

"Binary chemical cross-linking" is another new concept firstly introduced into the preparation of PVA-PAMPS membranes by our group [26,27]. It is a new composite made up of ether side chains that can be "ether-rich" like figure 14 shows. Different from the membrane preparation in acetone medium using GA as a cross-linker, chemically cross-linked PVA was designed on the basis of "binary cross-linking" processes, where bifunctional terephthalaldehyde (TA) was used as a main cross-linker o form the main chain and, the monofunctional polyalkylenes (two, four, and six CH2 groups) terminated by aldehydes were used as auxiliary cross-linkers to form the side chains. Thus, the resulted membranes attain both good mechanical property and excellent flexibility. This process is finished just by simply immersing the membrane samples in reaction reagents: n-butylaldehyde/terephthalaldehyde (BmTn), n-hexylaldehyde/terephthalaldehyde (HmTn), and n-octylaldehyde/terephthalaldehyde (OmTn) with N,N-dimethylformamide (DMF) as a reaction medium. The cross-linking proceeds between the hydroxyl groups (–OH) of PVA and the aldehyde groups (-CHO) of BmTn, HmTn, and OmTn in the membrane due to a "self-acid-catalyzed reaction" by PAMPS. Different compositions of binary cross-linking agents and different cross-linking times were employed to obtain membranes with different cross-linking densities and morphologies. Using this procedure, various membranes could be prepared according to the binary cross-linkers designated (table 4). For a comparison, by applying GA as a main cross-linker and n-octylaldehyde as an auxiliary cross-linker/or hydrophobicizer, different membrane samples were also prepared in different mass ratio of n-octylaldehyde/glutaraldehyde (OmGn), and they are all summarized in table 4.

Table 4a. Designation of PVA-PAMPS composites on the basis of binary chemical cross-linking (A): Terephthalaldehyde as a main cross-linker

Sample designation	Auxiliary cross-linkers/or Hydrophobicizer	Cross-linker
BmTn-system	n-Butylaldehyde (mass%)	Terephthalaldehyde (mass%)
B2T5	2	5
B5T5	5	5
B10T5	10	5
B20T5	20	5
HmTn-system	n-Hexylaldehyde (mass%)	Terephthalaldehyde (mass%)
H5T5	5	5
H10T5	10	5
H20T5	20	5
HmTn-system	n-Octylaldehyde (mass%)	Terephthalaldehyde (mass%)
O5T5	5	5
O10T5	10	5
O20T5	20	5

Table 4b. Designation of PVA-PAMPS composites on the basis of binary chemical cross-linking (B):Glutaraldehyde as a main cross-linker

Sample designation	auxiliary cross-linker/or Hydrophobicizer	Cross-linker
	n-Octylaldehyde (mass%)	Glutaraldehyde (mass%)
O10G1	10	1
O20G1	20	1
O30G1	30	1
O40G1	40	1

2.2.1. THERMAL PROPERTIES

Figure 15 shows the thermogravimetric analysis spectra of binary chemically cross-linked PVA-PAMPS with different contents of PAMPS, different spacer lengths of the CH2 group in the auxiliary cross-linkers, and different compositions of binary cross-linking agent. There are three major weight loss stages around 80-150, 200-330, and 400-450 °C followed by the final decomposition of the polymer blend. This result is similar to the plasticizer incorporated PVA-PAMPS semi-IPNs as we have described in section 2.1. For the samples being exposed to air prior to TG analysis, the weight losses in the first, second, and third stages can be attributed, respectively, to the expulsion of water molecules from polymer matrix or the moisture absorbed from the air, the decomposition of sulfonic acid groups (SO2 and SO3), and the splitting of the main chain of PVA followed by decomposition of the polymer backbone above 450°C. Although no dramatic change in TG profiles was observed for different chemical cross-linking compositions, earlier onset of the decomposition temperature was noted for the membranes with a higher content of PAMPS than those with a lower content of PAMPS (figure 15A). Earlier decomposition was also noted for the short spacer length of alkyl groups in auxiliary cross-linkers (figure 15B), following the order (CH2)2 > (CH2)4 > (CH2)6. This implies that a different inner morphology of the membranes is probably produced due to the modification of chemical cross-linking with the introduction of different auxiliary cross-linkers or "side chains".

The TG profiles by applying GA as a main cross-linker are showed in figure 15D for a comparison. Similarly, no dramatic change was observed by changing the OmGn compositions. Like the cases for TA main cross-linker, the first stage, occurring at 50–180oC with about 2–10% loss of initial weight, is considered to be the evaporation of residual water present in the polymer matrix. The highest weight loss of about 45% in the second and the third stage of decomposition started above 200oC and ended at around 380oC. This was followed by a final destruction of cross-linking bridges between 400 and 500oC with a further 15% weight loss. TG analysis, however, showed almost a two-stage decomposition process with O10G1 as a binary reaction agent, a three-stage decomposition with O20G1, and a four-stage decomposition by applying O30G1 and O40G1 as binary reaction agents, respectively. At the same time, the different degradation steps become more distinct indicating less drastic degradations. Nevertheless, the onset decomposition temperature of sulfonic acid groups seems to become evident when the mass ratio of the hydrophobicizer, octylaldehyde, to the main cross-linker, GA, increases.

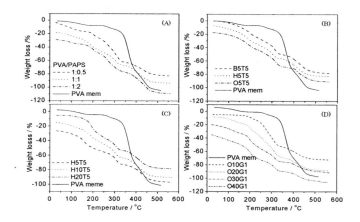

Figure 15. TG thermograms of pure PVA and binary chemically cross-linked PVA-PAMPS with different (A) content of PAMPS in polymer, (B) spacer length of the CH2 group in auxiliary cross-linkers, (C) composition of binary cross-linking agents and (D) binary reaction agent compositions with glutaraldehyde as a main cross-linker. (A) H5T5 was used as a binary cross-linking agent. (B, C) Polymer composition of PVA/PAMPS = 1:1 in mass. (D) Polymer composition of PVA/PAMPS =1:2 (in mass). The cross-linking time was 2h.

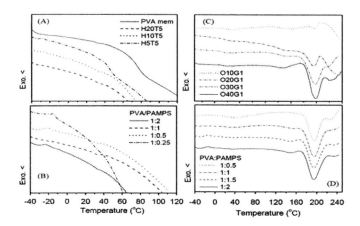

Figure 16. DSC thermograms of pure PVA and binary chemically cross-linked PVA-PAMPS in different (A) composition of binary cross-linking agents, (B) PAMPS content in polymer with TA as a main cross-linker and in different (C) composition of binary cross-linking agents, (D) PAMPS content in polymer with GA as a main cross-linker. (A) Polymer composition of PVA/PAMPS = 1:1 in mass. (B) H5T5 was used as a binary cross-linking agent. (C) Polymer composition of PVA/PAMPS = 1: 2 in mass. (D) O20G1 was used as a binary cross-linking agent. The cross-linking time was 2 h.

The results of DSC thermograms of the PVA–PAMPS composites modified on the basis of binary chemical cross-linking are plotted in figure 16. The glass transition temperature at about 108 °C of PAMPS was not observed [54]. For higher temperatures, the DSC traces exhibited a rapid downward shift indicating a fast endothermic process in the samples, which may be associated with the expulsion of water or the start of decomposition of PAMPS in the membrane as the TG profile shows in figure 15. A well-resolved step that is associated with the glass transition temperature, Tg, of pristine PVA was found to be at 78°C. This is in agreement with the results reported elsewhere [55,56]. Changing the binary cross-linking compositions, for example, lead to a reduction in Tg from 78°C for pristine PVA to 54°C by applying H20T5 as a cross-linking agent, and a further reduction to 34 °C by applying H10T5 and H5T5 as cross-linking agents, respectively (figure 16A). This indicates that the ordered association of PVA molecules was altered by the modification of chemical cross-linking. The addition of PAMPS lead to a decrease in glass transition temperature from 74 to 34 °C, but the effect of increasing PAMPS content in the membrane on Tg was negligible even at the highest acid content. The step change was too small to be established, and it even disappeared with a PVA/PAMPS mass ratio below 1:0.5 (figure 16B). Similarly, in the cases for GA main cross-linker system, the only phase transition observed in the -40 to 50oC temperature range. Changing the OmGn compositions presented no distinct differences in glass transition temperatures (figure 16C), and an increase of the PAMPS content produced similar Tg values at 30–40oC in the entire acid range (figure 16D). This is coincident with PVAH3PO2-H2O proton-conducting systems [41]. The disappearance of Tg and the exhibition of only a single glass transition temperature may suggest a well-blended PVA and PAMPS. In other words, the composites are in an amorphous, miscible state [41,57,58]. In addition, the endothermic and exothermic responses were observed around 200oC except for the expulsion of water molecules from the polymer matrices. Combined with the results of the TG provided, PVA-PAMPS composites are considered to be thermally stable up to 200oC in dry nitrogen.

2.2.2. SWELLING BEHAVIOR AND DIMENSIONAL CHANGE

Table 5 summarizes the physicochemical properties of various PVA-PAMPS composite membranes after modification on the basis of binary chemical cross-linking. The commercially available Nafion 117 membrane in its fully hydrated state is also listed in the table for a comparison. Note, the water uptake of Nafion

117 in its fully hydrated state ($n = 21$) agrees very well with the value reported elsewhere [59]. If the water uptake is defined as water molecules per sulfonic acid group, it was found that all the membrane samples reached high water uptakes, in excess of $n = 40$, regardless of different spacer lengths of the CH2 group in auxiliary cross-linkers or their compositions (table 5). Large water uptakes are probably because of the highly hydrophilic property of PAMPS or the amorphous structure of resulted membranes. The greater flexibility of the polymer chains (side chains) also could allow more water to reside between the polymer chains. This can be seen clearly from table 5, in which the swelling strongly increased with the mass ratio of auxiliary cross-linkers in the binary cross-linking agent. This, together with other polar groups present in the membranes leads to the formation of large ion clusters, which may allow larger sorption of water. High water uptake has also been found in other proton-conducting membrane systems, such as $n = 34$ in SPEEK [60], $n = 39$ in SPEEK/PBI [1], $n = 41$ in PVdF-g-PSSA [61], $n = 60$ in PVdF-SPS [62], $n = 47$ in PVdF-PS and even $n = 100$ in porous PVdF-HFP-PS [63] and $n = 176$ in PSU systems [64]. In general, water uptake dependencies of PVA-PAMPS are much greater than those of perfluorinated Nafion®, and they are strongly depended on temperature, humidity, content of PAMPS incorporated and composition of the binary cross-linking agent that is "side chain to main chain" as well as the spacer lengths of CH2 group in side chains.

In the cases for GA main cross-linker system, as seen from table 6, all the tested samples also reached high water uptakes, in excess of $\lambda = 39$. The membranes, which were prepared from the binary OmGn reaction agent with high mass content of octylaldehyde, e.g., O30G1 and O40G1, showed the largest absorption of water. The n-octylaldehyde functioned as a hydrophobicizer, is in fact acted as a 'side chain' through an acetal via reaction of PVA and n-octylaldehyde. Thus it may cause the greater flexibility of the polymer chains or the microstructure of the polymers, which leads to more water to reside between the polymer chains. However, the membranes showed a much lower swelling by using GA cross-linker as compared to those by using TA cross-linker, in which about 20% higher water uptake was obtained. This implies that the cross-linking density is strongly increased when the bifunctional GA, i.e., the aldehyde with alkyl chain as a cross-linker, is applied, although the mass content of n-octylaldehyde, is much higher than in TA main cross-linker system (see table 5 and 6). In spite of the relatively lower water uptakes of GA main cross-linker system, they showed high proton conductivities, which will be discussed further in the following section.

Table 5. Physicochemical properties of PVA-PAMPS composites on the basis of binary chemical cross-linking compared with those of Nafion 117

PVA/PAMPS ratio	Binary cross-linker	Time for cross-linking (h)	Water uptakea) (g g-1)	Freezing waterb) (wt%)	Bound water nH2O:-SO3H	IEC c) (meq. g-1)	σ / 25°C (S cm-1)
	BmTn-system						
1:1	B2T5	2	1.95	40(13)	53	1.63	0.12
PVA/PAMPS ratio	Binary cross-linker	Time for cross-linking (h)	Water uptakea) (g g-1)	Freezing waterb) (wt%)	Bound water nH2O:-SO3H	IEC c) (meq. g-1)	σ / 25°C (S cm-1)
	B5T5	2	2.34	44(15)	65	1.62	0.092
	B10T5	2	3.44	46(16)	105	1.58	0.088
	B20T5	2	5.61	76(27)	174	1.55	0.057
	HmTn-system						
1:1.5	H5T5	2	1.90	35(11)	49	1.66	0.110
	H10T5	2	2.36	43(14)	63	1.68	0.095
	H20T5	2	3.71	64(23)	113	1.51	0.078
	OmTn-system						
1:1.5	O5T5	2.5	1.57	38(13)	40	1.65	0.12
	O10T5	2.5	1.84	47(16)	47	1.61	0.10
	O20T5	2.5	3.12	50(19)	100	1.46	0.083
Nafion® 117 (fully hydrated)		–	0.34	8	13	0.91	0.1

Table 6. Water uptake, IEC and proton conductivity of PVA-PAMPS composites on the basis of binary chemical cross-linking

Polymer (PVA/PAMPS)	Composition of binary reaction agent	Time for cross-linking	WU (g g-1)	IEC (meq g-1)	σ /25°C (S cm-1)
	O10G1	2h	1.36	1.91	0.118
1:2	O20G1	2h	1.39	1.89	0.114
	O30G1	2h	1.57	1.75	0.094
	O40G1	2h	1.63	1.71	0.092

Dimensional changes (figure 17) of the membranes were investigated by immersion of the square pieces of samples into room temperature water for a day, which were then taken out and placed in an ambient condition (35% RH) for another 24 h. Changes of the ratio of the swollen length to the 'dry' length of the membranes were calculated by the following equation:

$$\Delta Lx = (Lx - Lxo)/Lxo$$
$$\Delta Ly = (Ly - Lyo)/Lyo \qquad\qquad (5)$$
$$\Delta Lz = (Lz - Lzo)/Lzo$$

where Lx, Ly, and Lz were the dimensions of the swollen membranes in length, width, and thickness, respectively, and the subscript o denoted the membranes equilibrated in an ambient condition (35% RH).

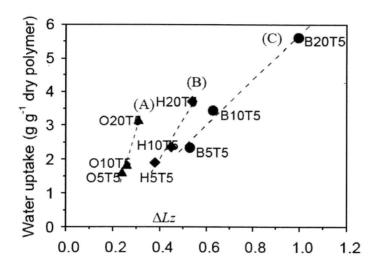

Figure 17. Dimensional changes (ΔLz) as a function of water uptake for binary chemically cross-linked PVA-PAMPS with different binary cross-linking agents and compositions. (A) Polymer composition of PVA/PAMPS = 1:1 in mass. (B,C) Polymer composition of PVA/PAMPS = 1:1.5 in mass. The cross-linking time was 2h.

The change in length due to swelling was found to be almost the same in all directions (in the Lx, Ly, and Lz directions), indicating an isotropic property of the chemically cross-linked PVA-PAMPS composites. However, strong swelling was observed for the membrane with the shortest CH2 chain (two CH2 groups) in the auxiliary cross-linker, which resulted in very large size changes. This is particularly evident at high PAMPS contents, where the increase in both thickness and surface area is easily reached. Contrary to this, the membranes with longer CH2 chains (six CH2 groups) showed a relatively lower swelling. The volume fractions thus calculated [65], of the water content in the swollen membranes, show corresponding results. This suggests that PVA-PAMPS having a shorter alkyl spacer is more sensitive to humidity than those having a longer alkyl spacer. Such a behavior is different from that of the covalently cross-linked PSUs, where the swelling increases with the chain length of the cross-linker [52]. This is an interesting result. For the auxiliary cross-linkers used in this work, only one end is cross-linked with PVA while the other end is free relative to the "double cross-linking" effect of TA or GA. These longer CH2 spacers play a "side chain" role and twine around each other to form an imaginary hydrophobic part, thus restraining the excessive swelling of a polymer (see figure 14). As seen in table 7, the change in length due to swelling was found to be almost the same in all directions also for GA main cross-linker system, indicating further a rather isotropic swelling of PVA–PAMPS membranes. The volume fractions, Φ, thus calculated from eqn.[65]:

$$\Phi = (LxLyLz - LxoLyoLzo)/LxLyLz \qquad (6)$$

are almost the same (around 0.56) regardless of the compositions of binary reaction agents. On the other hand, Nafion 117 showed high proton conductivity at a relatively lower Φ, suggesting an ion-rich channels in Nafion® membrane.

Table 7. Water volume fraction, Φ, dimensional change, ΔL, and the state of water of PVA-PAMPS composites on the basis of binary chemical cross-linking compared with those of Nafion 117

Polymer (PVA/PAMPS)	Composition of binary reaction agent	Φ	ΔLx	ΔLy	ΔLz	Freezing water[a] λ	Bound water λ
	O10G1	0.57	0.35	0.35	0.28	7.3	32.3
1:2	O20G1	0.55	0.36	0.33	0.27	10.2	30.7
	O30G1	0.56	0.37	0.36	0.28	17.9	31.9
	O40G1	0.56	0.37	0.31	0.26	20	32.9
Nafion 117 (fully hydrated)		0.38	0.15	0.17	0.20	8	13

a) Calculated from the DSC analyses. Here, the heat of fusion of pure ice (334 J/g) was used for evaluating the degree of crystallinity from the melt endotherm in DSC thermograms. λ denotes the number of water molecules per a sulfonic acid group.

2.2.3. THE STATE OF WATER

The water uptake of membranes is a key issue to the fuel cell performances. However, water sorption characteristics are of great importance for water-swollen polymer membranes. This is because materials containing a large amount of water in their networks show their characteristic physicochemical properties depending not only on the water content but also on the states of water. For a proton-conducting fuel cell membrane, in fact, that transport strongly depends on the nature of water, i.e., the state of water rather than the degree of water sorption [66]. Therefore, the water sorption behavior in swelled membranes has become one of the essential factors. The state of water is usually categorized into the following three different types: (i) free water, which shows the same temperature and enthalpy of melting/crystallization as bulk water; (ii) freezing bound water, weakly bound to the ionic and polar groups in the polymer matrix, which exhibits a melting/crystallization temperature shifted with respect to that of bulk water but it can be detected by melting transitions in DSC measurements and (iii) nonfreezing water, strongly bound to the ionic and polar parts of the polymer, which shows no detectable phase transition over the range of temperatures normally associated with bulk water.

Figure 18 gives the typical DSC heating traces of binary chemically cross-linked PVA-PAMPS composites after being fully swollen. The broad bimodal melting peaks with a good resolution were clearly observed from -15 to 5oC in the tested samples. These peaks were assigned to the freezable water (freezing bound water and free water) [34,67]. Samples with a high content of sulfonic acid groups (PVA/PAMPS = 1:1.5 in mass) showed both a melting onset and a melting point, -5.2 and 3.25 °C, whereas only a melting point of 3.27 °C was observed in the samples with a low content of sulfonic acid groups (PVA/PAMPS = 1:0.5 in mass) (figure 18A). Evidently, the number of sulfonic groups influences the crystallization and melting temperatures [34,68]. We noted that the membrane samples with O5T5 as a binary cross-linking agent clearly showed both a lower melting onset and a lower melting point as compared to samples with O10T5 and O20T5 as binary cross-linking agents (figure 18B). Also, a lower melting onset and a lower melting point were observed in samples with longer CH2 spacers in the auxiliary cross-linkers, as compared to samples with shorter CH2 spacers in the auxiliary cross-linkers (figure 18C). This implies that increased cross-linking density leads to lower crystallization and melting temperatures because of its incorporated sulfonic acid groups with respect to the water content [34,68]. Similar freezing peaks were also observed in the membranes with a water melting point around 0°C [69,70] However, it produced only a negligible influence due to its very small melting peak.

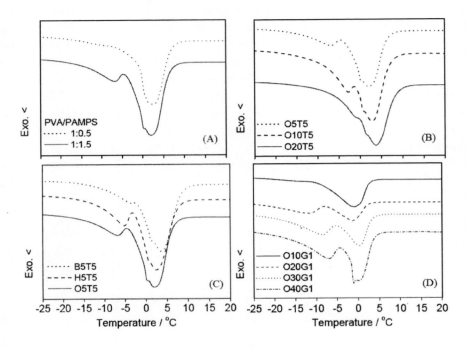

Figure 18. DSC heating traces of binary chemically cross-linked PVA-PAMPS in different (A) content of PAMPS in polymer, (B) composition of binary cross-linking agents, (C) spacer length of the CH2 group in auxiliary cross-linkers with TA as a main cross-linker and (D) composition of binary cross-linking agents with GA as a main cross-linker. (A) O5T5 was used as a binary cross-linking agent. (B,C) Polymer composition of PVA/PAMPS = 1:1.5 in mass. (D) Polymer composition of PVA/PAMPS = 1:2 in mass. The cross-linking time was 2 h.

For OmGn binary cross-linking system (that is with GA as a main cross-linker), the samples with low mass content of hydrophobicizer, octylaldehyde (O10G1 and O20G1), clearly showed a lower melting onset as compared to those with a high mass content of hydrophobicizer (O30G1 and O40G1) like the cases for TA main cross-linker system (figure 18D). This implies that polymer modification by increased hydrophobicizer i.e., high content of octylaldehyde leads to a higher crystallization and melting temperature of water because of the amorphous structure of the polymer that may be induced by hydrophobicizer [34,68]. It is worth noticing that in comparison with the freezing water content, the membranes prepared all showed a large sorption of nonfreezing water (table 5 and 7). Water molecules bound strongly to ionic species and polar sites in the polymer chain. As one knows the chemically cross-linked PVA–PAMPS polymer composites contain other polar parts such as the amide groups –C–NH, the free – CHO groups and even the –OH groups from non-reacted PVA main chains. There may be enough of a driving force at room temperature to force water into PVA–PAMPS chains and thus swell the polymer, besides the strong interaction between the water and the sulfonic acid groups [67,68,71]. The amorphous morphology of the membranes as DSC traces showed may be another reason for the high non-freezing water uptake [60,68]. It was found that the high methanol permeability of Nafion membranes is due to their high freezing water uptake compared to other polymer membranes in spite of the fact that the total water uptake was lower [66]. Therefore, larger sorption of non-freezing water may be an attractive membrane characteristic when they are intended for use in DMFCs. The electroosmotic drag of the water, and thus the methanol permeability, may be low due to the strong association of the water to the ionomer [66-68].

2.2.4. PROTON CONDUCTIVITY AND ION-EXCHANGE CAPACITY (IEC)

Figure 19 presents the corresponding proton conductivities (σ) of PVA-PAMPS composites in different composition of binary cross-linking agents (figure 19A), and different spacer lengths of the CH2 group in side chains (figure 19B) as a function of the PAMPS content in the membrane with TA as a main cross-linker. As expected, the σ values increased with the content of PAMPS in the polymer. All the membrane samples offered high proton conductivity that is comparable to that of Nafion 117 except for the B20T5- and H20T5-based ones (figure 19 and table 5). It is clear that the content of hydrophilic components and their distribution in the matrix of the membrane play a key role in proton conduction. Especially at high PAMPS contents, a connection of a hydrophilic network will build up in the membrane, which allows a fast exchange between the mobile ions and the fixed sites through the hydrated hydrophilic network [72]. The maximum of proton conductivity reaching up to 0.12 S cm-1 was obtained for a polymer composition of PVA/PAMPS of 1:1 in mass with B2T5 as a binary cross-linking agent and 1:1.5 in mass using H5T5 and O5T5 as binary cross-linking agents. Additional higher content of PAMPS, for example, when the PAMPS/PVA mass ratio is above 1.5, causes a decrease in proton conductivity (figure 19A) and the similar trends were also for the spacer length dependence of the conductivity (figure 19B).

By applying GA as a main cross-linker, changing the binary OmGn reaction agent compositions gives similar proton conduction behavior, that is, with an increase in PAMPS content, the conductivity increased significantly and reached a plateau in a polymer composition of PVA/PAMPS at 1:2 in mass (figure 19C). All of the tested samples displayed σ values 0.09–0.118 S cm-1, compared with Nafion 117, which provided 0.091 S cm-1 in its fully hydrated state. Similarly, additional higher PAMPS content in the membrane lead to a leveled off of the conductivity, for example, when the mass ratio of PVA/PAMPS was above 1:2.

Generally, proton conductivity will increase with an increase in WU because a larger free volume contributes to the high mobility of free ions. Also, the proton conductivity increases with an increase in IEC because of the high charge density of the membranes [38]. For clarifying the mechanism of the proton conduction in binary chemically cross-linked PVAPAMPS, figure 20 illustrates the proton conductivity together with the WU and IEC of PVA-PAMPS membranes plotted against the content of PAMPS. We can see that the IEC showed a formal linear increase with the PAMPS content in polymer, whereas the WU shows an exponential increase with the PAMPS content in the polymer (figure 20A). This suggests that the proton conductivity is subject more to the swelling effect than to the IEC. A further increase in the amount of water at higher PAMPS content does not contribute to proton conductivity, but rather a dilution of charge carriers. This effect is particularly evident for the binary cross-linking compositions B20T5, H20T5, and O20T5, where the membranes show excessive swelling and, the water uptake reaches as high as ~300% with respect to their initial mass (see table 5). Contrary to this, the O-side chain based system displayed higher proton conductivity despite its relatively lower water uptake than B- and H-based cross-linking ones, although they have the similar IECs (table 5). This is probably due to the longer hydrophobic side chains of the O-based system as described previously. They could form a better microphase separation structure, where the ionic domains are connected to each other, forming the proton-conducting pathways. As to this point, it could be demonstrated much better for GA main cross-linker system in which n-octylaldehyde also was used as auxiliary cross-linker like the cases for TA main cross-linker system. As figure 20B shows, the water uptake exhibited an exponential relationship vs. IEC at a further high content of PAMPS in polymer (the insert). In other words, too much water uptake inversely affects the conduction. Thus the energy barrier to the proton transport increases, which leads to a plateau in the proton conductivity (figure 19).

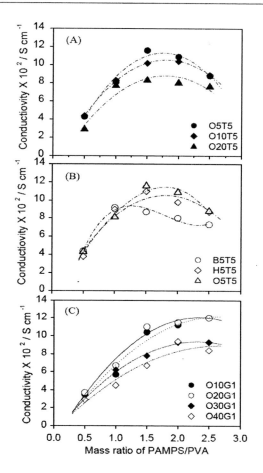

Figure 19. Proton conductivity changes as a function of the PAMPS/PVA mass ratio for binary chemically cross-linked PVA-PAMPS in different (A) composition of binary cross-linking agents, (B) spacer length of the CH2 group in auxiliary cross-linkers with TA as a main cross-linker, and (C) composition of binary cross-linking agents with GA as a main cross-linker. The cross-linking time was 2 h. The values were obtained at 25 ± 2°C.

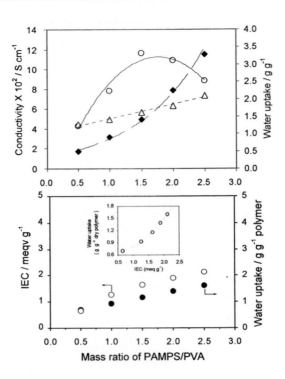

Figure 20. (A) Proton conductivity (o), IEC (Δ), and water uptake (□) change as a function of the PAMPS/PVA mass ratio for binary chemically cross-linked PVA-PAMPS. O5T5 as a binary reaction agent, TA as a main cross-linker. (B) Water uptake (●) and IEC (o) plotted as a function of PAMPS/PVA mass ratio for binary chemically cross-linked PVA-PAMPS. O20G1 as a binary reaction agent, GA as a main cross-linker. The values were obtained at 25 ± 2 oC.

The temperature dependences of proton conductivity for various PVA-PAMPS composites are compared in figure 21. The polymer composition of PVA-PAMPS was fixed at 1:1 in mass with B5T5 as a binary cross-linking agent and 1:1.5 in mass with H5T5 and O5T5 as binary cross-linking agents (TA as a main cross-linker). All the membrane samples exhibited positive temperature-conductivity dependencies, i.e., follows the Arrhenius relationship $\sigma = \sigma_0 \exp(-E_a/kT)$, in the tested temperature range from 5 to 50 °C, whether they were prepared with different binary cross-linking compositions (figure 21A) or with different spacer lengths of the CH2 group in the auxiliary cross-linkers (figure 21B). Similarly, all the membrane samples exhibit positive temperature–conductivity dependencies regardless of the binary OmGn compositions or

PAMPS contents in the polymer by applying GA as a main cross-linker instead of TA (figure 21C and21D). The membranes showed the highest proton conductivity of 0.14 S cm-1 at 43 ± 2oC or 0.098 S cm-1 at 5 ±2oC by applying O5T5 as a binary cross-linking agent and, of 0.14 S cm-1 at 43 ± 2oC or 0.083S cm-1 at 5 ±2oC when O10G1 and O20G1 as binary reaction agents were applied. The apparent activation energy values fall into the range of 6.3 - 11.9 kJ mol-1 derived from the slope of log σ vs 1/T plots for TA main-cross linker system, 9–13 kJ mol-1 for GA main cross-linker system and, increased slightly with an increase in the mass ratio of the auxiliary cross-linker (hydrophobicizer or side chain) to the main cross-linker (main chain). In brief, these are a little lower than the Ea of Nafion 117, for which Ea = 14.6 kJ mol-1. It can be deduced, therefore, that both the vehicle mechanism and the Grotthuss mechanism are responsible for the proton conduction in the present system.

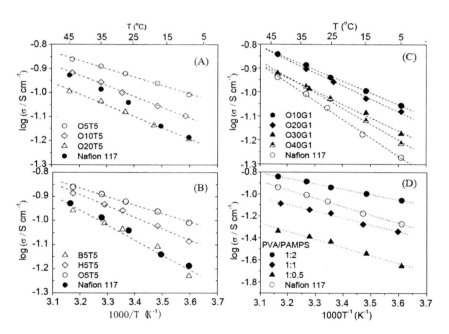

Figure 21. Temperature dependences of proton conductivity of binary chemically cross-linked PVA–PAMPS in different (A) composition of binary reaction agents, (B) spacer length of the CH2 group in auxiliary cross-linkers with TA as a main cross-linker and (C) composition of binary reaction agents, (D) content of PAMPS in polymer with GA as a main cross-linker. Polymer composition: PVA/PAMPS = (A) 1:1.5 in mass, (B) 1:1 in mass with B5T5 as a binary reaction agent and 1:1.5 in mass with O5T5- and H5T5 as a binary reaction agent, (C) 1:2 in mass with O10G1 as a binary reaction agent. The cross-linking time was 2 h.

2.2.5. METHANOL PERMEABILITY

As described above, a proton exchange membrane should serve several functions besides its high proton conductivity, such as good water stability, reasonable thermal stability and appropriate mechanical property. It should have even lower methanol permeability if methanol is used as fuels of PEM fuel cells. Table 8 illustrates the methanol permeability (PM) and the selectivity (σ/PM) of some typical PVA-PAMPS composites on the basis of binary chemical cross-linking. It can be seen that Nafion membrane displayed the highest methanol permeability of the membranes tested. PVA-PAMPS membranes based on O10G1 and O20G1 showed the lowest methanol permeability, and their PM values were about one half of those of Nafion 117. As seen from table 5 and 6, the PVA-PAMPS membranes showed much larger water sorption behaviour than Nafion 117. The PVA-PAMPS membrane sample prepared from O40G1 binary reaction agent even showed about 5 times larger WU value than Nafion 117, but its PM was still lower than Nafion 117 including the membrane prepared from O5T5. This indicates that methanol cross-over can be efficiently suppressed by the effect of cross-linking and hydrophobic modification. Higher methanol permeability of Nafion®could be explained by the higher fraction of freeing water (loosely bound and free water) [65-67,73-75] Contrary to this, PVA-PAMPA composites share large sorption of non-freezing water relative to the freezing water (table 5 and 7), thus hindering the methanol transportation. Therefore, the methanol permeability could be reduced by controlling the free water in the membranes. Due to the appreciably lower methanol permeability of PVA-PAMPS composites compared to Nafion 117, higher selectivity (σ/PM) was thus attained for PVA-PAMPS membranes, which is near 2 fold higher than Nafion 117 by applying O10G1 and O20G1 as binary reaction agents.

Table 8. Methanol permeability (PM) and selectivity (σ/ PM) of PVA-PAMPS composites on the basis of binary chemical cross-linking

PVA/PAMPS (mass ratio)	Composition of binary reaction agent	σ (25oC)	PM ($\times$ 106 cm2 s-1)	σ/ PM ($\times$ 104 S cm-3 s)
1:2	O10G1	0.118	1.29	9.2
1:2	O20G1	0.114	1.41	8.1
1:2	O30G1	0.094	1.65	5.7
1:2	O40G1	0.093	1.8	5.1
1 :1.5	O5T5	0.12	1.8	6.7
Nafion117		0.091	2.13	4.3

2.2.6. MECHANICAL BEHAVIOR

The mechanical properties of binary chemically cross-linked PVA–PAMPS composites were tested for the tensile strength and tensile elongation. Some typical membrane candidates prepared with OmGn binary reaction agent are given in table 9. All tested samples are ductile, for example, at 22±2oC and 45% RH. It can be seen that both the tensile strength and the elongation increased when increasing the mass ratio of the hydrophobicizer, n-octylaldehyde, but the reverse was not so, indicating the miscibility between the components. This strength may be accounted for, in large part, by constraints posed by entanglements formed between the bifunctional GA and the long side chains of n-octylaldehyde during the cross-linking reaction. Similar trends in terms of the modulus to Nafion 117 were even observed (table 8). Although less elongation and greater property changes of PVA-PAMPS composites after they are immersed in pure water, the PVA–PAMPS composites showed a good mechanical property owing to the high flexibility of the side chains provided by the hydrophobicizer, which allows easy relaxation of the polymer chains.

Table 9. Mechanical properties of PVA-PAMPS composites on the basis of binary chemical cross-linking compared with those of Nafion 117 at 22±2oC and 45% RH

Polymer (PVA/P AMPS)	Composition of binary reaction agent	Time for cross-linking	Young's modulus (MPa %-1)	Tensile strength (Kg mm-2)	Elongation (%)
	O10G1	2h	21.3	1.3	7.2
1:2	O20G1	2h	20.5	1.8	12.6
	O30G1	2h	27.1	2.2	17.8
	O40G1	2h	28.7	2.2	16.2
Nafion 117		–	23.8	2.4	11.8

2.2.7. HYDROLYTIC AND OXIDATIVE STABILITY AND THE MEMBRANES MODIFIED WITH HIGH MOLECULAR WEIGHT PVA

Time dependent measurements of the proton conductivity were carried out at 25oC, and at an elevated temperature of 50oC, respectively, to evaluate the stability of binary chemically cross-linked PVA–PAMPS. Generally, membranes with high IEC values tend to yield large water uptakes and thus a poor water stability. However, all the membranes presented here showed excellent water stability for a long time without any decrease in proton conductivity (e.g., 25 oC for 240 h) and also a stable proton conductivity with time measured at 50oC for the polymer composition PVA/PAMPS = 1 : 2 in mass [27].That is, the membranes not only showed an excellent water stability but also a good thermal stability (e.g., 50oC) due to the modification by both the chemical cross-linking and the side chain hydrophobicizer, although their WU and IEC were both much higher than that of Nafion 117.

All the membranes also showed good water stability for more than one month after immersion in 50°C water for TA main cross-linker system. Further boiling experiment in 100oC showed that the membrane sample prepared from B-side chain was dissolved completely after boiling in 100oC water for 2 hours. The membrane sample prepared from H-side chain did not dissolved but showed an excessive swelling after boiling in 100oC water for 2 hours. However, the membranes prepared from O5T5 and O10T5 binary reaction agents showed much better hydrolytic stability without any changes in appearance, flexibility, and toughness even after being boiled in 100 °C water for 24 h [26]. This is in a well agreement with the results obtained with OmGn binary cross-linking system. As we mentioned above, the swelling increases with the chain length of the cross-linker for covalently cross-linked PSUs [52]. But the opposite trend was observed for binary chemically cross-linked PVA-PAMPS composites, that is, the membranes swelling decreased with longer CH2 chains, in which the membrane prepared from O-side chain (six CH2 groups) shows the lowest swelling effect. These longer CH2 "side chain" could twine around each other to form an imaginary hydrophobic part, thus restraining the excessive swelling of a polymer. Also, the membrane stability is highly improved. Therefore, this innovative design based on binary chemical cross-linking is expected to improve the performances of the PVA-PAMPS family membranes. It will also be applicable to other polymer systems to be used as electrolytes in electrochemical cells with high proton conductivity.

Table 10. Water stability of PVA-PAMPS composites on the basis of binary chemical cross-linking

PVA/PAMPS (mass ratio)	Composition of binary reaction agent	Time for cross-linking	50oC (in water)	100oC (boiling water)
1:1	B5T5	2h	168h, brittle	2h, soluble in water
1:1.5	H5T5	2h	o	10h, excessive swelling
1 :1.5	O5T5	2.5h	o	o
1 :1.5	O10T5	2.5h	o	o
1 :1.5	O20T5	2.5h	o	o

The oxidative stability of PVA-PAMPS composites was tested by subjecting to Fenton reagent in an aqueous solution of $H2O2(3\%)/FeSO4(2ppm)$ at a fixed treatment temperature of 60°C, where the membrane stability toward the oxidative effect was tracked as weight loss as a function of immersion time. As figure 22 shows, almost no weight loss of the membranes was observed within initial 2-3 hours regardless of different binary cross-linking reaction agents, then a sharp decrease in the membrane weight occurred within another 2-3 hours. But after that a weight about 30-40 wt% of the originals was retained and reached an equilibrium with no longer weight loss occurring, again. Different from the cases for plasticizer incorporated PVA-PAMPS, PVA-PAMPS composites on the basis of binary chemical cross-linking did not dissolved into Fenton's reagent even at high treating temperature of 60oC for more than 300 hours. This is the first time to be found that a fully hydrocarbon membrane did not dissolve into Fenton reagent at all.

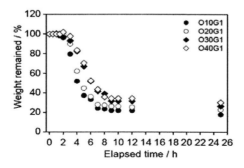

Figure 22. Oxidative stability of binary chemically cross-linked PVA-PAMPS composites in H2O2(3%)/FeSO4(2ppm) at 60oC. OmGn as a binary cross-linking reaction agent. Polymer composition: PVA/PAMPS = 1:2 in mass.

Table 11. PVA/PAMPS conducting composite using high molecular weight PVA (HMw-PVA)

Sample		Cross-linker		
		n-Octylaldehyde (%)	Glutaraldehyde (%)	Cross-linking time (h) at 25oC in DMF
HMw-PVA30	O5G1	5	1	2
	O10G1	10		
	O20G1	20		
	O30G1	30		
PVA-OKS	O5G1	5	1	2
	O10G1	10		
	O20G1	20		
	O30G1	30		
HMw-PVA40	O5G1	5	1	2
	O10G1	10		
	O20G1	20		
	O30G1	30		

Recently, novel PVA with high molecular weight (HMw-PVA), which is called HMw-PVA30 (HMw-PVA (OH 99.8mol%, Vis. 137.7mPa· s, AcONa 0.1wt%, DP = 3120), OKS-PVA (OKS-PVA (AA-PVA; OH 99.3mol%, Vis. 56.6mPa· s, AcONa0.05%, DP = 2200) containing –OCOCH2COCH3) and HMw-PVA40 (HMw-PVA (OH 99.7mol%, Vis. 296.6mPa· s, AcONa 1wt%, DP = 4090), has been successfully suggested as ingredients [32]. By using HMw-PVA as polymer matrices, the membrane performances are highly improved due to the improving of micro-phase structure in the membranes. With OmGn as binary cross-linking agent, for example, HMw-PVA/PAMPS membranes showed high proton conductivities around 0.06 to 0.09 S cm-2 (figure 23A). The mass ratio of HMw-PVA to PAMPS was 1:1, which is lower than in the membranes prepared from low molecular weight PVA (LMw-PVA, OH 99 mol%, Viscosity 66 mPa s) [26,27], but high proton conductivities were retained with lower water uptakes (around 0.7 – 0.9 g g-1 dry polymer depending on different cross-linking compositions) (figure 23B). It seems that the micro-structure becomes denser due to modification of HMw-PVA, thus the more hydrophobic property of the membranes emerged leading to decrease in water uptake. Low methanol permeability (PM) that was 1/3 - 1/5 of Nafion 117 was afforded as figure 24 shows. In particular, the HMw-PVA/PAMPS membranes exhibited a very good longevity, for example, by subjecting the membranes to Fenton reagent in an aqueous solution of H2O2(3%)/FeSO4(2ppm) at a treatment temperature of 60°C.

They could withstand the processing in Fenton's reagent for six hours for HMw-PVA30 and OKS-PVA based composite, and eight hours for HMw-PVA40 based composite before the observable weight loss occurred (figure 25). Similarly, these PVA/PAMPS composites did not dissolve ultimately into Fenton's reagent at 60oC regardless the immersion time. This is very different from other proposed membranes with aromatic skeletons like we mentioned in previous sections [46,49,50,51]. Considering the excellent water stability of PVA-PAMPS toward water (e.g., at 50oC) and their good mechanical strength and flexibility as described above, by using high molecular weight PVA, the PVA-PAMPS composite membranes is expected to be promising characteristics, especially for applications to DMFCs.

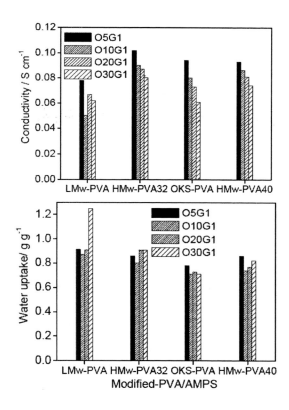

Figure 23. (A) Proton conductivity and (B) water uptake of HMw-PVA/PAMPS composites with different cross-linking compositions. Polymer composition: HMw-PVA/PAMPS = 1:1 in mass.

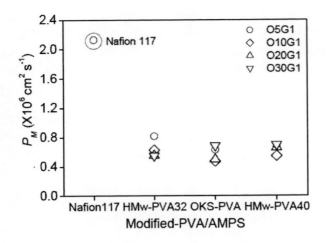

Figure 24. Methanol permeability of HMw-PVA/PAMPS composites with different cross-linking compositions. Polymer composition: HMw-PVA/PAMPS = 1:1 in mass.

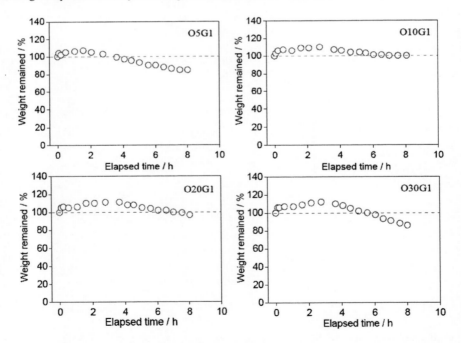

Figure 25. Oxidative durability of HMw-PVA40/PAMPS composites with different cross-linking compositions. Polymer composition: HMw-PVA40/PAMPS = 1:1 in mass.

Chapter 5

CELL PERFORMANCES OF PVA-PAMPS HYDROCARBON PROTON-CONDUCTING ELECTROLYTE MEMBRANES IN A REAL DMFC

2.3.1. NAFION SOLUTION AS BOTH CATALYTIC INK AND BINDER

The cell performances in the cell configuration: 10% MeOH | Pt-Ru/C | PVA-PAMPS-R | Pt/C | O2 being used in a single-cell of DMFC were tested with PVA-PAMPS-R (R = PEGDCE, PEG, PEGME, PEGDE and PEGBCME) fabricated membrane electrode assemblies (MEAs). MEA was prepared by hot-pressing the catalyst loaded carbon paper (Toray TGP-H-090) to the PVA-PAMPS-R membranes at the pressure of 100 kg cm-2 at 110oC for 5 minutes. The active electrode area for a single cell test was 4 cm2. The catalyst loadings on the anode (by a rolling method) and the cathode electrodes were 2 mg(Pt-Ru) cm-2 (30%Pt-15Ru%/C, Johnson-Matthey HiSPEC 7000) and 1mg(Pt) cm-2 (ElectroChem), respectively. Nafion polymer loadings (5% Nafion solution used as the binder) on both electrodes were 0.7-1.1 mg cm-2. MEA was then inserted into a fuel cell hardware which consisted of graphite block with machined serpentine flow channel and copper current collectors (figure 26). 10 wt% aqueous solution of methanol was pumped into the anode channel of the cell under atmospheric pressure and, pure oxygen as the cathode fuel was supplied to the cathode channel with a gas flow rate 100 mL min-1 through humidifiers under ambient pressure. The cell was placed in a temperature-controlled chamber, which was used to keep the cell at a constant temperature. Polarization curves were obtained using a fuel cell evaluation system (TFC-2100, Sokken) over the temperature range of 25 – 80oC.

A   B   C D E F G   Cell assembly

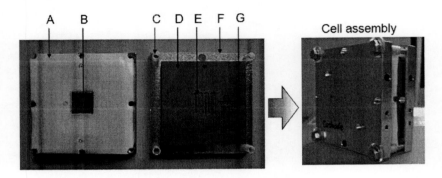

Figure 26. Photograph of DMFC showing (A) Teflon sheets, (B) MEA, (C) screw, (D) graphite block, (E) gas channel, (F) current collector and (G) reference electrode compartment.

Figure 27 shows I–V characteristics of DMFC with PVA-PAMPS-R semi-IPN proton-conducting membranes by incorporating different plasticizer variants. All the PVA-PAMPS semi-IPN membranes could withstand the stress encountered during MEA assembling in the fuel cell test although they showed much higher water-uptake that that of Naion 117 and, no obvious flowerage rimples were observed. As seen in figure 27, all the membrane samples showed good cell performances with the open circuit potential (OCP) values around 0.7 – 0.73 V. This is similar or even a little higher than that of Nafion 115. In addition, the OCP values have no obvious changes regardless of the temperature or the content of PAMPS in membranes, indicating the low methanol permeability characteristic of PVA-PAMPS semi-IPNs. An increase in the fuel cell temperature leads to a dramatic improvement in the cell performance. In spite of a low catalyst loading (2 mg cm−2 on the anode and 1 mg cm-2 on the cathode), high power densities were obtained from 17.2 mW cm-2, 20.8 mW cm-2, 17.6 mW cm-2 and 20.3 mW cm-2 at 25oC to 49.8 mW cm-2, 52.3 mW cm-2, 44.5 mW cm-2 and 48.1 at 80oC, by applying PEGDCE, PEG, PEGME and PEGDE as plasticizer, respectively. It seems that PEGDCE and PEG incorporated the PVA-PAMPS semi-IPNs show the best cell performances. However we can not give an assert on this issuer considering some errors for each experiment. In fact, all the membranes exhibit good performances, say, transparency, uniform, good mechanical and flexibility, if the preparation conditions are well controlled. Note, since the oxygen humidifier was maintained just at room temperature, this suggests that PVA-PAMPS-R semi-IPNs could work in DMFC without or with room temperature humidification of the cathode reactant, allowing simplified auxiliary systems.

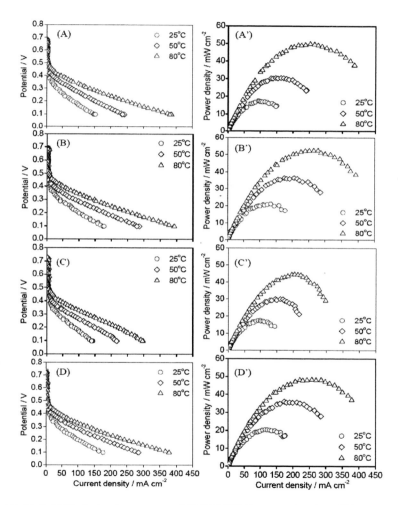

Figure 27. (A,B,C,D) plarization curve and (A',B',C',D') power density of PVA-PAMPS-R semi-IPNs in DMFC mode at different temperatures. Nafion solution was used as both the catalyst ink and the binder. R = (A,A') PEGDCE, (B,B') PEG, (C,C') PEGME and (D,D') PEGDE. Catalyst loading on the anode is Pt-Ru 2mg cm-2, and Pt 1mg cm-2 on the cathode. Polymer composition: PVA/PAMPS/PEGDCE = 1:1:0.5, PVA/PAMPS/PEG = 1:1:0.5, PVA/PAMPS/PEGME = 1:1:0.4 and PVA/PAMPS/PEGDE = 1:1:0.3 in mass, respectively.

2.3.2. PVA/PAMPS SOLUTION AS BOTH CATALYTIC INK AND BINDER

MEA is often the "heart" of a polymer electrolyte membrane fuel cells, since the power performance essentially depends on the quality of the MEA. The conversion efficiency of MEA depends on many factors such as the membrane material, nature of the catalyst and the binder used in the electrodes, their ratio and finally the method of MEA preparation. For a good contact between the membranes (usually as an electrolyte) and the catalyst coated carbon paper (usually as electrodes) in MEA, the use of the catalyst ink and binders which are similar in chemical nature to the membrane, is strongly suggested [76]. The solubilized Nafion used in these electrodes is often not compatible with the new polymers, and as a result, cell performance decreased drastically [46]. Unfortunately, most of the work carried out on MEA fabrication is still restricted to Nafion solution because of no specific method available up to now. On the basis of this situation, we attempted to prepare MEAs with PVA-PAMPS-R proton-conducting semi-IPN membranes as electrolyte, and by applying PAMPS, PVA/PAMPS and PVA/PAMPS/PEGDCE as both the catalyst ink and the binder to improve adhesion between the electrolyte and catalyst. The following four different strategies were adopted to fabricate MEAs: i) the catalyst ink was made from a solution of 5 wt.% Nafion; ii) the catalyst ink was made from a solution of 2 wt.% PAMPS dissolved in EtOH; iii) the catalyst ink was made from a solution of PVA/PAMPS (0.5:1 wt.%) dissolved in 60 wt.% EtOH and, iv) the catalyst ink was made from a solution of PVA/PAMPS/PEGDCE (0.5:1:0.5 wt.% and 0.5:1:2 wt.%) dissolved in 60 wt.% EtOH. The above catalyst inks (ii-iv) were also used for the corresponding binders for the anode, and Nafion solution was employed for the cathode during MEA fabrications.

The cell performance was found to be highly improved by applying PAMPS as both the catalyst ink and the binder. The MEAs fabricated with 2% PAMPS showed a power density of 14 mW cm-2 at 25o and increased to 29 mW cm-2 at 80oC with a anode catalyst loading just only 0.6 mg cm−2. Evidently, this can be attributed to the high proton-conducting characteristic of PAMPS, with compatibility between the impregnation solution PAMPS (precursor sol) and the membrane. Unfortunately, the interface stability of MEA was not ideal, and delamination of electrodes from the catalyst layer was found during the running of the cell because of the strong hydrophilic property of PAMPS. Increasing the hot-pressing temperature up to 120oC did not prompt the complete self cross-linking of PAMPS.

A good trapping of PAMPS was achieved, for example, by adding slightly PVA and PVA/PEGDCE into PAMPS solution during hot-pressing treatment. Figure 28 illustrates the I-V and performance results of MEA fabricated from above procedures. It could be noted that MEA fabricated using PVA/PAMPS as both the catalyst ink and the binder (figure 28A and A'), showed a good performance that was similar to MEA fabricated with Nafion solution (figure 27A and A'). Around 210 mV, the current density reached around 154, 231 and 348 mA cm-2 at 25, 50 and 80oC, respectively as compared with the MEAs fabricated with Nafion solution at 25oC (141 mA cm-2), 50oC (213 mA cm-2) and 80oC (324 mA cm-2), respectively. This may due to the proton-conducting characteristic of PAMPS together with the chemical similarity between impregnation solution PVA-PAMPS and membrane, which improves good adhesion of electrode and catalyst. Out of our anticipation, MEAs fabricated with PVA/PAMPS/PEGDCE solution exhibited a lower cell performance than MEAs fabricated with either PVA/PAMPS solution or Nafion solution (figure 28B and B'; figure 28C and C'). The gap of cell performance becomes larger with increasing temperature, regardless of different weight ratio of PEGDCE in the binders, in spite of the most similar chemical structure to PVA-PAMPS-PEGDCE semi-IPN membrane.

Under the same measuring conditions (except that MEA was fabricated at 135oC and 100 kg cm-2 for 3 min), the MEA fabricated with Nafion 115 showed a power density of 75 mW cm-2 at 80oC. In spite of the comparable proton conductivity to Nafion, the PVA-PAMPS-PEGDCE semi-IPN shows a relatively lower cell performance in a real DMFC mode, especially at high temperatures. This may be due to a good contact of Nafion membrane with both the catalyst and the gas diffusion electrode (GDE), and therefore a perfect proton transfer between Nafion membrane and Nafion-based catalyst layers. Formation of a good three-phase boundary at the catalyst layer and at the GDE/membrane interface is very important to attain high performance of a fuel cell. Another important factor may be a much higher water uptake of PVA-PAMPS-PEGDCE semi-IPN than Nafion membranes. PVA and PEGDCE both have hydrophilic property, and this gives rise to a poorer membrane electrode interface due to the swelling of the membrane at high temperatures especially for high content of PEGDCE, thus a relatively lower cell performance. However, these initial results are still promising, since the PVA-PAMPS-R semi-IPNs are all hydrocarbon chains, which are very different from the perfluorosulfonic membranes such as Nafion®, or other reported membranes with aromatic skeletons, just as we stated previously. The MEAs using PVA-PAMPS proton-conducting semi-IPN membranes showed much better cell performances than those using other membranes like polyimide [46] (with a metal loading on the anode 10 mg cm-2), sPEEK [77] (with a metal loading on the anode 4 mg cm-2) and SEBS [78] (with a metal loading on the anode 3 mg cm-2). Therefore, by increasing the catalyst loading as employed in other reports (4 – 10 mg cm-2) [46,77-79], in this case, PVA-PAMPS-R proton-conducting semi-IPNs would prove to be good candidates when they are intended for use in low temperature DMFCs.

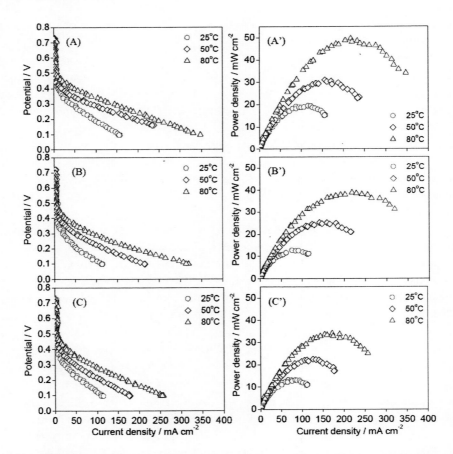

Figure 28. (A,B,C) Polarization curve and (A',B',C') power density of PVA-PAMPS-PEGDCE semi-IPN in DMFC mode at different temperatures. (A, A') PVA/PAMPS (0.5:1 wt.%) was used as both the catalyst ink and the binder; (B,B') PVA/PAMPS/PEGDCE (0.5:1:0.5 wt.%) was used as both the catalyst ink and the binder; (C,C') PVA/PAMPS/PEGDCE (0.5:1:2 wt.%) was used as both the catalyst ink and the binder. Catalyst loading on the anode is Pt-Ru 2mg cm-2, and Pt 1mg cm-2 on the cathode. Polymer composition: PVA-PAMPS-PEGDCE = 1:1:0.5 in mass.

2.3.3. LIFE TIME OF DMFC

Considering the high proton conductivity but relative a lower water uptake, the long-term DMFC test was conducted using PVA-PAMPS-PEGBCME fabricated MEA for a polymer composition PVA/PAMPS/PEGBCME = 1:0.75:0.4 in mass as a typical candidate at 50oC, by applying a constant voltage 0.25 V through the cell. The time course of the cell current density and the OCV were tracked during a daily start daily stop mode operation, and the results are given in figure 29. The OCV showed values of 0.68 to 0.73 V, and the cell current density at 0.25 V was between 41 and 120 mA cm-2 for the initial two days, and after that reached a stable value around 60 mA cm-2 with OCV 0.70V. A striking feature of the initial long-term test is that no appreciable decay of the current density was observed during the whole operation time long than130 hours, and so was the power density (figure 30). The high stability of cell performance may be due to the suppressed methanol permeability owing to the effective cross-linking in the membrane and its stability by a special plasticizer effect of PEGBCME. Such a long-term test of DMFC is an interesting result considering that PVA-PAMPS composite are all hydrocarbon membranes made merely of aliphatic skeletons. Although it has not reached the cell performances yet like some other good candidates such as the sulfonated polysulfone (which reached 3000h fixed at 0.5V at 80oC) and acid–base blend polymer polysulfone-2-amide-benzimidazole/sulfonated poly(ether ether ketone) (which reached 120h at 80oC), all with sulfonated aromatic membranes or prepared by introducing fluorine groups [80,81], or for Nafion® membranes [82,83], this affords the PVA-PAMPS composites a unique position compared to most of the proposed membranes for DMFC applications.

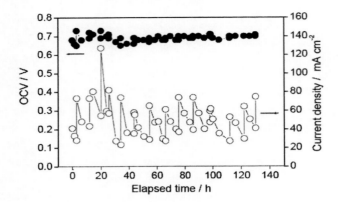

Figure 29. Variations of OCV and current density with operating time using PVA-PAMPS-PEGBCME semi-IPN at 50oC. Polymer composition PVA/PAMP/PEGBCME = 1:0.75:0.4 in mass. Catalyst loading on the anode is Pt-Ru 1.9mg (Pt 1.23mg) cm-2, and Pt 1mg cm-2 on the cathode. Methanol concentration and flow rate are 10 wt% and 5ml min-1, respectively. O2 was supplied to the fuel cell with a flow rate 100ml min-1, through the humidifier at room temperature.

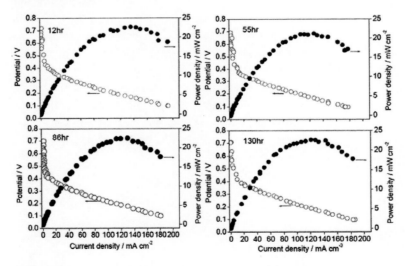

Figure 30. Variations of polarization curve and power density with operating time using PVA-PAMPS-PEGBCME semi-IPN in DMFC at 50oC. Polymer composition PVA/PAMP/PEGBCME = 1:0.75:0.4 in mass. Catalyst loading on the anode is Pt-Ru 1.9mg(Pt 1.23mg) cm-2, and Pt 1mg cm-2 on the cathode. Methanol concentration and flow rate are 10 wt% and 5ml min-1, respectively. O2 was supplied to the fuel cell with a flow rate 100ml min-1, through the humidifier at room temperature.

Chapter 6

PVA-BASED OTHER POLYMER MEMBRANES AND THEIR APPLICATION FOR PEFC AND AFC

It should be mentioned that polymer blend is attracting intense interest in recent years owing to its potential feature of combining the good characteristics of each blend component while at the same time reducing their deficient characteristics [84]. Not only from the viewpoint of cost effective materials and improved toughness of the membranes, but also the synergistic effects are important considerations that are not equal to the 'physical' average. Terpolymers is a preferred route for reaching this aim [85]. A typical candidate is PVA-PAMPS-PVP [28], where poly(vinylpyrrolidone) (PVP), as a stabilizer, was successfully introduced into PVA-PAMPS in combination with chemical cross-linking. By blending PVA as third component, the membrane performances, such as the methanol permeability, mechanical property and the oxidative stability were largely improved along with its high proton conductivity [28]. Another one is cross-linked PVA membranes containing sulfonic acid group, in which sulfosuccinic acid (SSA) is used both as a cross-linking agent and a proton conductor, therefore provides the PVA membrane good proton conductivity along with low methanol permeability [73-75]. By applying the above procedure, a high power density of 0.7 W cm^{-2} was achieved when they were used for PEFC with cross-linked poly(vinyl alcohol) and poly(styrene sulfonic acid-co-maleic anhydride) as proton-exchange fuel cell membranes [86]. Partially phosphorylated PVA also was suggested as novel ionic conducting-membranes [87].

Most recently, PVA/SSA and PVA/TiO$_2$ membranes have been reported for use in alkaline fuel cells (AFC) by applying methanol as anode fuels [88,89].

Although the maximum power density was only about 4.13 mWcm^{-2} and 7.54 mWcm^{-2} at 60°C under 1 atm in 2M KOH+2MCH$_3$OH solution, the use of less expensive non-precious metal catalysts, such as Ag, Ni and MnO$_2$ ensures no doubt the promising advantages over other fuel cells. Therefore, we can see that there are still a lot of posibilities to improve the performance of PVA-based membranes not limited on DMFC only, but it can also be extend for other fields of fuel cells like PEFC and AFC.

ACKNOWLEDGEMENT

The authors would like to give their appreciations to Dr. Takeo Hamaya for his close ontribution and devotion to this work from the early stage of the project.

REFERENCES

[1] Kreuer, KD. On the development of proton conducting polymer membranes for hydrogen and methanol fuel cells. *J. Membr. Sci.,* 2001, 185, 29–39.

[2] N.,Vladimir; Martin, J; Wang, H.; Zhang, J. A review of polymer electrolyte membranes for direct methanol fuel cells. J. Power Sources, in Press, Available online 24 March 2007

[3] Lamy, C; L´eger, JM; Srinivasan, S. in: J.O'M. Bockris, B.E. Conway, R.E. White (Eds.), Modern Aspects of Electrochemistry, 34, Kluwer Academic, New York, 2001 (Chapter 3).

[4] Dyer, CK. Fuel cells for portable applications. *J. Power Sources,* 2002, 106, 31–34.

[5] Wasmus, S; Kuver, A. Methanol oxidation and direct methanol fuel cells: a selective review. *J. Electroanal. Chem.,* 1999, 461, 14–31.

[6] Costamagna, P; Srinivasan, S. Quantum jumps in the PEMFC science and technology from the 1960s to the year 2000: Part I. Fundamental scientific aspects. *J. Power Sources,* 2001, 102, 242–252.

[7] Heinzel, A; Barragan, VM. A review of the state-of-the-art of the methanol crossover in direct methanol fuel cells. *J. Power Sources,* 1999, 84, 70–74.

[8] Qi, Z; Hollett, M; He, C; Attia, A; Kaufman, A. Operation of Direct *Methanol Fuel Cells. Solid-State Lett.,* 2003, 6, A27–A29.

[9] Holmberg, S; Holmlund, P; Wilén, CE; Kallio, T; Sundholm, G; Sundholm, F. Synthesis of proton-conducting membranes by the utilization of preirradiation grafting and atom transfer radical polymerization techniques. *J. Polymer Sci., Part A: Polym. Chem.,* 2002, 40, 591–600.

[10] Li, Q; He, R; Jensen, JO; Bjerrum, N. J. Approaches and Recent Development of Polymer Electrolyte Membranes for Fuel Cells Operating above 00 °C. *Chem. Mater.* 2003, 15, 4896– 4915.

[11] Inzelt, G; Pineri, M; Schultze, JW; Vorotyntsev, MA. Electron and proton conducting polymers: recent developments and prospects. *Electrochim. Acta,* 2000, 45, 2403–2421.

[12] Rikukawa, M; Sanui, K. Proton-conducting polymer electrolyte membranes based on hydrocarbon polymers. *Prog. Polym. Sci.,* 2000, 25, 1463-1502.

[13] Jones, DJ; Rozière, J. Recent advances in the functionalisation of polybenzimidazole and polyetherketone for fuel cell applications. *J. Membr. Sci.,* 2001, 185, 41–58.

[14] Kerres, JA. Development of ionomer membranes for fuel cells. *J. Membr. Sci.,* 2001, 185, 3–27.

[15] Zhang, J; Xie, Z; Zhang, J; Tang, Y; Song, C; Navessin, T; Shi, Z; Song, D; Wang, H; Wilkinson, DP; Liu, ZS; Holdcroft, S. High temperature PEM fuel cells. *J. Power Sources,* 2006, 160, 872–891.

[16] Jochen, MH; Taeger, A; Vogel, C; Schlenstedt, K; Lenk, W; Lehmann, D. Membranes from sulfonated block copolymers for use in fuel cells. *Sep. Purif. Technol,* 2005, 41, 207–220.

[17] Park, JS; Park, JW; Ruckenstein, E. A dynamic mechanical and thermal analysis of unplasticized and plasticized poly(vinyl alcohol)/methylcellulose blends. *J. Appl. Polym. Sci.,* 2001, 80, 1825–1834

[18] Chuang, WY; Young, TH; Chiu, WY; Lin, CY. The effect of polymeric additives on the structure and permeability of poly(vinyl alcohol) asymmetric membranes. *Polymer,* 2000, 41, 5633–5641.

[19] Nagura, M; Hamano, T; Ishikawa, H. Structure of poly(vinyl alcohol) hydrogel prepared by repeated freezing and melting. *Polymer,* 1989, 30, 762–765

[20] Rhim, JW; Yeom, CK; Kim, SW. Modification of poly(vinyl alcohol) membranes using sulfur-succinic acid and its application to pervaporation separation of water-alcohol mixtures. *J. Appl. Polym. Sci.,* 1998, 68, 1717–1723.

[21] Chiang, WY; Chen, CL. Separation of water—alcohol mixture by using polymer membranes—6. Water—alcohol pervaporation through terpolymer of PVA grafted with hydrazine reacted SMA. *Polymer,* 1998, 39, 2227–2233.

[22] Rhim, JW; Kim,YK. Pervaporation separation of MTBE-methanol mixtures using cross-linked PVA membranes. *J. Appl. Polym. Sci.,* 2000, 75, 1699–1707.

[23] Rhim, JW; Sohn, MY; Joo, HJ; Lee, KH. Pervaporation separation of binary organic-aqueous liquid mixtures using crosslinked PVA membranes. I.

Characterization of the reaction between PVA and PAA. *J. Appl. Polym. Sci.,* 1993, 50, 679–684.

[24] Pivovar, BS, Wang,Y; Cussler, EL. Pervaporation membranes in direct methanol fuel cells. *J. Membr. Sci.,* 1999, 154, 155–162.

[25] Alberti, G; Casciol, M. Solid state protonic conductors, present main applications and future prospects. *Solid State Ionics,* 2001, 145, 3–16.

[26] Qiao, JL; Hamaya, T; Okada, T. Chemically modified poly(vinyl alcohol)/2-acrylamido-2-methyl-1-propanesulfonic acid (PVA-PAMPS) as novel proton-conducting fuel cell membranes. *Chem. Mater.,* 2005, 17, 2413–2421.

[27] Qiao, JL; Hamaya, T; Okada, T. New highly proton conductive polymer membranes poly(vinyl alcohol)/2-acrylamido-2-methyl-1-propanesulfonic acid (PVA-PAMPS). *J. Mater. Chem.,* 2005, 15, 4414–4423.

[28] Qiao, JL; Hamaya, T; Okada, T. New highly proton-conducting membrane poly(vinylpyrrolidone)(PVP) modified poly(vinyl alcohol)/2-acrylamido-2-methyl-1-propanesulfonic acid (PVA-PAMPS) for low temperature direct methanol fuel cells (DMFCs). *Polymer,* 2005, 46, 10809–10816.

[29] Hamaya, T; Inoue, S; Qiao, JL; Okada, T. Novel proton-conducting polymer electrolyte membranes based on PVA/PAMPS/PEG400 blend. *J. Power Sources,* 2006, 156, 311–314.

[30] Qiao, JL; Okada, T. Highly durable proton conducting semi-IPNs from PVA/PAMPS composites by incorporating plasticizer variants. Electrochem. *Solid State Lett.,* 2006, 9, A379–381.

[31] Qiao, JL; Ikesaka, S; Saito, M; Kuwano, J; Okada, T. New binders for MEA fabrication for low temperature DMFCs using PVA-PAMPS proton-conducting semi-IPN membranes. *Electrochemistry,* 2007, 75, 126–12.

[32] Qiao, JL; Ono, H; Oishi, T; Okada, T. Performance Enhancement of PVA/PAMPS conducting composites using high molecular weight PVA. *ECS Transactions,* 2006, 3, 97–102.

[33] Qiao, JL; Ikesaka, S; Saito, M; Kuwano, J; Okada, T. Life test of DMFC using poly(ethylene glycol) bis(carboxymethyl)ether plasticized PVA/PAMPS proton conducting semi-IPNs. *Electrochem. Comm.,* 2007, 9, 1945-1959.

[34] Karlsson, LE; Wesslen, B; Jannasch, P. Water absorption and proton conductivity of sulfonated acrylamide copolymers. *Electrochim. Acta,* 2002, 47, 3269–3275.

[35] Zukowska, G; Williams, J; Stevens, JR; Jeffrey, KR; Lewera, A; Kulesza, PJ. The application of acrylic monomers with acidic groups to the synthesis of proton-conducting polymer gels. *Solid State Ionics,* 2004, 167, 123–130.

[36] Su, PG; Tsai, WY. Humidity sensing and electrical properties of a composite material of nano-sized SiO2 and poly(2-acrylamido-2-methylpropane sulfonate). *Sens. Actuators B,* 2004, 100 , 417–422

[37] Walker Jr, CW. Proton-conducting polymer membrane comprised of a copolymer of 2-acrylamido-2-methylpropanesulfonic acid and 2-hydroxyethyl methacrylate. *J. Power Sources,* 2002, 110, 144–151

[38] Kang, MS; Choi,YJ; Moon, SH. Water-swollen cation-exchange membranes prepared using poly(vinyl alcohol) (PVA)/poly(styrene sulfonic acid-co-maleic acid) (PSSA–MA). *J. Membr. Sci.,* 2002, 207, 157–170.

[39] Kang, MS; Kim, JH; Won, JW; Moon, SH; Kang, YS. Highly charged proton exchange membranes prepared by using water soluble polymer blends for fuel cells. *J. Membr. Sci.,* 2005, 247, 127–135.

[40] Panero, S; Fiorenza, P; Navarra, MA; Romanowska, J; Scrosati, B. Silica-added, composite Poly(vinyl alcohol) membranes for fuel cell application. *J. Electrochem. Soc.,* 2005, 152, A2400–A2405.

[41] Vargas, RA; Zapata, VH; Matallana, E; Vargas, MA. More thermal studies on the PVOH/H3PO2/H2O solid proton conductor gels. *Electrochim. Acta,* 2001, 46, 1699–1702.

[42] Awadhia, A; Patel, SK; Agrawal, SL. Dielectric investigations in PVA based gel electrolytes. *Prog. Cryst. Growth Charact. Mat.,* 2006, 52, 61–68

[43] Shapiro, YE; Shapiro, TI. 1H NMR self-diffusion study of PVA cryogels containing ethylene glycol and its oligomers. *J. Colloid and Interface Sci.,* 1999, 217, 322–327

[44] Mansur, HS; Orefice, RL; Mansur, AAP. Characterization of poly(vinyl lcohol)/poly(ethylene glycol) hydrogels and PVA-derived hybrids by small-angle X-ray scattering and FTIR spectroscopy. *Polymer,* 2004, 45, 7193–7202.

[45] M. Saito, N. Arimura, K. Hayamizu, T. Okada, Mechanisms of ion and water transport in perfluorosulfonated ionomer membranes for fuel cells. *J. Phys. Chem. B,* 2004, 108, 16064–16070.

[46] Einsla, BR; Kim, YS; Hickner, MA; Hong, YT; Hill, ML; Pivovar, BS; McGrath, JE. Sulfonated naphthalene dianhydride based polyimide copolymers for proton-exchange-membrane fuel cells: II. Membrane properties and fuel cell performance. *J. Membr. Sci.,* 2005, 255, 141–148.

[47] Sulfonated poly(arylene ether sulfone) copolymer proton exchange membranes: composition and morphology effects on the methanol permeability. *J. Membr. Sci.,* 2004, 243, 317–326.

[48] Savadogo, O. Emerging membranes for electrochemical systems: Part II. High temperature composite membranes for polymer electrolyte fuel cell (PEFC) applications. *J. Power Sources,* 2004, 127, 135–161.

[49] Yasuda, T; Miyatake, K; Hirai, M; Nanasawa, M; Watanabe, M. Synthesis and properties of polyimide ionomers containing sulfoalkoxy and fluorenyl groups. *J. Polymer Sci., Part A: Polym. Chem.,* 2005, 43, 4439 - 4445.

[50] Kim, DS; Guiver, MD; Nam, SY; Yun, TI; Seo, MY; Kim, SJ; Hwang, HS; Rhim, JW; Preparation of ion exchange membranes for fuel cell based on crosslinked poly(vinyl alcohol) with poly(styrene sulfonic acid-co-maleic acid). *J. Membr. Sci.,* 2006, 281, 156–162.

[51] Kerres, J; Tang, CM; Graf, C. Improvement of properties of poly(ether ketone) ionomer membranes by blending and cross-linking. *Ind. Eng. Chem. Res.,* 2004, 43,4571–4579.

[52] Koter, S; Piotrowski, P; Kerres, J. Comparative investigations of ion-exchange membranes. *J. Membr. Sci.,* 1999, 153, 83–90.

[53] Park, JS; Park, JW; Ruckenstein, E. Thermal and dynamic mechanical analysis of PVA/MC blend hydrogels. *Polymer,* 2001, 42, 4271–4280.

[54] Ding, J; Chuy, C; Holdcroft, S. A self-organized network of nanochannels enhances Ion conductivity through polymer films. *Chem. Mater.,* 2001, 13, 2231–2233.

[55] Petty-Weeks, S; Polak, AJ. Differential scanning calorimetry and complex admittance analysis of PVA/H3PO4 proton conducting polymer blends. *Sens. Actuators,* 1987, 11, 377–386.

[56] Vargas, MA; Vargas, RA; Mellander, BE. More studies on the PVAl+H3PO2+H2O proton conductor gels. *Electrochim. Acta,* 2000, 45, 1399–1403.

[57] Smitha, B; Sridhar, S; Khan, AA. Synthesis and characterization of proton conducting polymer membranes for fuel cells. *J. Membr. Sci.,* 2003, 225, 63–76.

[58] Li, ZF; Ruckenstein, E. Improved surface properties of polyaniline films by blending with Pluronic polymers without the modification of the other characteristics. *J. Colloid Interface Sci.,* 2003, 264, 362–369.

[59] Zawodzinski, TA; Springer, TE; Davey, J; Jestel, R; Lopez, C; Valerio, J; Gottesfeld, S. A Comparative Study of Water Uptake By and Transport Through Ionomeric Fuel Cell Membranes. *J. Electrochem. Soc.,* 1993, 140, 1981–1985.

[60] Kopitzke, RW; Linkous, CA; Anderson, HR; Nelson, GL. Conductivity and Water Uptake of Aromatic-Based Proton Exchange Membrane Electrolytes. *J. Electrochem. Soc.,* 1993, 147, 1677–1681.

[61] Hietala, S; Skou, E; Sundholm, F. Gas permeation properties of radiation grafted and ulfonated poly-(vinylidene fluoride) membranes. *Polymer,* 1999, 40, 5567–5573.

[62] Ostrovskii, DI; Torell, LM; Paronen, M; Hietala, S; Sundholm, F. Water sorption properties of and the state of water in PVDF-based proton conducting membranes. *Solid State Ionics, 1997,* 97, 315–321.

[63] Soresi, B; Quartarone, E; Mustarelli, P; Magistris, A; Chiodelli, G. PVDF and P(VDF-HFP)-based proton exchange membranes. *Solid State Ionics.* 2004, 166, 383–389.

[64] Kerres, JA. Development of ionomer membranes for fuel cells. *J. Membr. Sci.,* 2001, 185, 3–27.

[65] Yeom, CK; Lee, KH. Vapor permeation of ethanol-water mixtures using sodium alginate membranes with crosslinking gradient structure. *J. Membr. Sci.,* 1997, 135, 225–235.

[66] Kim, YS; Dong, L; Hickner, MA; Glass, TE; Webb, V; McGrath, JE. State of Water in disulfonated poly(arylene ether sulfone) copolymers and a perfluorosulfonic acid copolymer (Nafion) and its effect on physical and electrochemical properties. *Macromolecules,* 2003, 36, 6281–6285.

[67] Cho, KY; Eom, JY; Jung, HY; Choi, NS; Lee, YM; Park, JK; Choi, JH; Park, KW; Sung, YE. Characteristics of PVdF copolymer/Nafion blend membrane for direct methanol fuel cell (DMFC). *Electrochim. Acta,* 2004, 50, 583–588.

[68] Karlsson, LE; Jannasch, P. Polysulfone ionomers for proton-conducting fuel cell membranes: sulfoalkylated polysulfones. *J. Membr. Sci.,* 2004, 230, 61–70.

[69] Baba, T; Sakamoto, R; Shibukawa, M; Oguma, K. Solute retention and the states of water in polyethylene glycol and poly(vinyl alcohol) gels. *J. Chromatogr. A,* 2004, 1040, 45–51.

[70] Kim, SJ; Park, SJ; Kim, SI. React. Synthesis and characteristics of interpenetrating polymer network hydrogels composed of poly(vinyl alcohol) and poly(N-isopropylacrylamide). *Funct. Polym,* 2003, 55, 61–67.

[71] Soresi, B; Quartarone, E; Mustarelli, P; Magistris, A; Chiodelli, G. PVDF and P(VDF-HFP)-based proton exchange membranes. *Solid State Ionics,* 2004, 166, 383–389.

[72]   Sumner, JJ; Creager, SE; Ma, JJ; Desmarteau, DD. Proton Conductivity in
       Nation117 and in a Novel Bis[(perfluoroalkyl)sulfonyl]imide Ionomer
       Membrane. *J. Electrochem. Soc.*, 1998, 145, 107–110.

[73]   Rhim, JW; Park, HB; Lee, CS; Jun, J H; Kim, DS; Lee,YM. Crosslinked
       poly(vinyl alcohol) membranes containing sulfonic acid group: proton and
       methanol transport through membranes. *J. Membr. Sci.*, 2004, 238, 143–
       151.

[74]   Kim, DS; Park, HB; Rhim, JW; Lee, YM. Preparation and characterization
       of crosslinked PVA/SiO2 hybrid membranes containing sulfonic acid
       groups for direct methanol fuel cell applications. *J. Membr. Sci.*, 2004, 240,
       37–48.

[75]   Kim, DS; Park, HB; Rhim, JW; Lee,YM. Proton conductivity and methanol
       transport behavior of cross-linked PVA/PAA/silica hybrid membranes.
       Solid State Ionics, 2005, 176, 117–126.

[76]   Thangamuthu, R; Lin, CW. Membrane electrode assemblies based on sol–
       gel hybrid membranes — A preliminary investigation on fabrication
       aspects. *J. Power Sources*, 2005,150, 48–56.

[77]   Li, L; Zhang, J; Wang,Y. Sulfonated poly(ether ether ketone) membranes
       for direct methanol fuel cell. *J. Membr. Sci.*, 2003, 226, 159–167.

[78]   Jung, DH; Myoung, YB; Cho, SY; Shin, DR; Peck, DH. A performance
       evaluation of direct methanol fuel cell using impregnated tetraethyl-
       orthosilicate in cross-linked polymer embrane. *International J. Hydrogen
       Energy*, 2001, 26, 1263–1269.

[79]   Lin, CW; Thangamuthu, R; Yang, C. J. Proton-conducting membranes with
       high selectivity from phosphotungstic acid-doped poly(vinyl alcohol) for
       DMFC applications. *J. Membr. Sci.*, 2005, 253, 23–31.

[80]   Kim,YS; Pivovar, B. Durability of Membrane-Electrode Interface under
       DMFC Operating Conditions. *ECS Transactions*, 2005,1(8), 457–467.

[81]   Fu, Y; Manthiram, A; Guiver, MD. Acid–base blend membranes based on
       2-amino-benzimidazole and sulfonated poly(ether ether ketone) for direct
       methanol fuel cells. *Electrochem. Comm.*, 2007, 9, 905–910.

[82]   Knights, SD; Colbow, KM; Pierre, JS; Wilkinson, DP. Aging mechanisms
       and lifetime of PEFC and DMFC. *J. Power Sources*, 2004, 127, 127–134.

[83]   Kim, H; Shin, SJ; Park,Yg; Song, J; Kim, Ht. Determination of DMFC
       deterioration during long-term operation. *J. Power Sources*, 2006, 160,
       440–445.

[84]   Manea, C; Mulder, M. Characterization of polymer blends of
       polyethersulfone/sulfonated polysulfone and polyethersulfone/sulfonated

polyetheretherketone for direct methanol fuel cell applications. *J Membr. Sci.*, 2002, 206, 443–453.

[85] Kudva, RA; Keskkula, H; Paul, DR. Compatibilization of nylon 6/ABS blend susing glycidyl methacrylate/methyl methacrylate copolymers. *Polymer,* 1998, 39, 2447–2460.

[86] Lin,CW; Huang,YF; Kannan, AM. Semi-interpenetrating network based on cross-linked poly(vinyl alcohol) and poly(styrene sulfonic acid-co-maleic anhydride) as proton exchange fuel cell membranes. *J. Power Sources,* 2007, 164 , 449–456.

[87] Takada, N; Koyama,T; Suzuki, M; Kimura, M; Hanabusa, K; Shirai, H; Miyata, S. Ionic conduction of novel polymer composite films based on partially phosphorylated poly(vinyl alcohol). *Polymer,* 2002, 43, 2031–2037.

[88] Yang, CC; Chiu, SJ; Chien, WC. Development of alkaline direct methanol fuel cells based on crosslinked PVA polymer membranes. *J. Power Sources,* 2006, 162, 21–29.

[89] Yang, CC. Synthesis and characterization of the cross-linked PVA/TiO2 composite polymer membrane for alkaline DMFC. *J. Membr. Sci.,* 2007, 288, 51–60.

INDEX

D